Discovering Engineering Design in the 21st Century

An Activities-Based Approach

Bradley A. Striebig
Professor of Engineering
James Madison University

Australia • Brazil • Canada • Mexico • Singapore • United Kingdom • United States

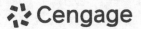

Discovering Engineering Design in the 21st Century: An Activities-Based Approach, **First Edition**
Bradley A. Striebig

SVP, Product: Cheryl Costantini

VP, Product: Thais Alencar

Portfolio Product Director: Rita Lombard

Senior Portfolio Product Manager:
Timothy L. Anderson

Product Assistant: Emily Smith

Learning Designer: MariCarmen Constable

Content Manager: Alexander Sham

Associate Digital Project Manager: John Smigielski

VP, Product Marketing: Jason Sakos

Director, Product Marketing: Danae April

Content Acquisition Analyst: Deanna Ettinger

Production Service: Lumina Datamatics Ltd.

Designer: Gaby McCracken

Cover Image Source: Bradley A. Striebig

For product information and technology assistance, contact us at
Cengage Customer & Sales Support, 1-800-354-9706
or **support.cengage.com.**

For permission to use material from this text or product, submit all requests online at **www.copyright.com.**

Library of Congress Control Number: 2023903489

Student Edition:
ISBN: 978-0-357-68520-4

Loose-leaf Edition:
ISBN: 978-0-357-68536-5

Cengage
200 Pier 4 Boulevard
Boston, MA 02210
USA

Cengage is a leading provider of customized learning solutions. Our employees reside in nearly 40 different countries and serve digital learners in 165 countries around the world. Find your local representative at **www.cengage.com.**

To learn more about Cengage platforms and services, register or access your online learning solution, or purchase materials for your course, visit **www.cengage.com.**

Printed at CLDPC, USA, 05-23

Contents

Preface

Discovering Engineering Design in the 21st Century: An Activities-Based Approach, First Edition, is a practical and applied introduction to the engineering needs of today's world. Engineering educators have learned that many students, especially engineering students, learn best when relating their own experience to the practice and knowledge base that forms the foundation of engineering education. The author has developed a text that provides foundational knowledge, traditional engineering skills, and hands-on experiential learning activities for a novel approach to introductory engineering courses. This curriculum-based text is focused on applying engineering principles to real-world design and problem analysis. It includes specific step-by-step examples and case studies for solving complex conceptual and design problems in several different engineering fields. This textbook also applies the principles of sustainable design to issues in both developed and developing countries. Students will benefit from having a companion guidebook as they begin their exploration into the engineering profession.

Dr. John Crittenden, Eminent Scholar at Georgia Tech, has suggested that engineering solutions include the following important elements/steps: (1) translating and understanding societal needs into engineering solutions such as infrastructures, products, practices, and processes; (2) explaining to society the long-term consequences of these engineering solutions; and (3) educating the next generation of scientists and engineers to acquire both the depth and the breadth of skills necessary to address the important physical and behavioral science elements of environmental problems and to develop and use integrative analysis methods to identify and design sustainable products and systems. Students will practice these steps throughout each chapter of this textbook as they engage with the issues facing this generation of engineers.

Organization and Potential Syllabus Topics

In *Discovering Engineering Design in the 21st Century: An Activities-Based Approach*, students are encouraged to explore and investigate problem-solving approaches through the engineering design process. The text is designed to engage students and encourage them to apply principles of math and science to solve problems. The engineering design approach is encouraged through activities designed to connect elements of visual drawing, an organized approach to utilize math and science principles, and consideration of safety and error analysis as a part of the engineering design process.

This text uniquely demonstrates how engineers use static information (books and equations), online resources (cost information, databases, reports, and so on), numerical analysis, and engineering critical thinking skills to synthesize solutions to today's problems. It provides resources for hands-on learning activities that engage students in identifying, formulating, and solving open-ended engineering problems. Problems are contextualized and related to professional engineering fields of practice and multiple engineering majors (see the table below) to help students produce solutions that consider public health, safety, and welfare. The open-ended exercises encourage students to analyze and interpret data, design and conduct experiments, and develop engineering judgment to draw conclusions. Rubrics are provided in the accompanying instructor's manual to help students understand approaches to engineering analysis and design as well as to evaluate students' work with respect to global, cultural, social, environmental, and economic factors.

Chapters and their relationship to traditional fields in engineering

Chapter Number	Chapter Title	Traditional Engineering Field(s)
1	Engineering Concepts and Simple Tools: Materials, Mass, Gravity, and Moment Arms	Engineering science and engineering technology
2	Measurements and Experiments: Data and Decisions with Sensors, Radios, and Arduinos	Electrical and computer engineering
3	Structures and Society: Structural Engineering Problem-Solving Techniques to Design Long-Lasting Solutions	Civil and structural engineering
4	Sustainable Development Goals: Engineering for Environmental Sustainability	Civil and environmental engineering
5	Food, Water, and Nutrients in Chesapeake Bay: An Earth Systems Engineering Approach	Agricultural, biological, and systems engineering
6	Approaches to Engineering Design	Industrial engineering
7	Designing for Energy Efficiency	Mechanical and petroleum engineering
8	Life Cycle Thinking: Understanding the Complexity of Sustainability	Industrial, materials, and process engineering

Supplements

Additional instructor resources for this product are available online. Instructor assets include a Solution Answer Guide, Image Libraries, and PowerPoint® slides. Sign up or sign in at www.cengage.com to search for and access this product and its online resources.

Acknowledgments

The author is grateful to his colleagues and students for the contributions they have made to the development of this textbook. Although there are too many contributors to name, a few deserve special mention. The faculty and staff in the Department of Engineering at James Madison University will recognize contributions to this work for which Dr. Striebig is indebted. He is particularly grateful for the contributions to the textbook from Dr. Daniel Castaneda, Dr. Jason Forsyth, and Dr. Shraddha Joshi. He would also like to recognize Jasmine White, Ayana Oancea, and Erica Dobies for their assistance in reviewing and preparing the text. The author would like to acknowledge John Wild, Dr. Robert Prins, Dr. Robert Nagel, and Dr. S. Keith Holland for assisting with the development of the drawings, concepts, and approaches for teaching first-year engineering students. Dr. Striebig also appreciates the efforts of Andrew Sklavounos, Hunter Goodman, Jack Peot, Jacob Messner, Lacey Monger, and Tyrees Swift-Josey for their contributions to Chapter 4.

The Chapter 5 content is built on the description of Earth systems engineering as described by Michael E. Gorman, who developed an Earth systems engineering course in the Department of Systems Engineering at the University of Virginia and concepts promoted by industrial ecologist Braden Allenby. Other influential work that contributed to the curriculum includes the case study on the Florida Everglades Restoration Project; the NASA Earth Science Enterprise Plan; course and curricula descriptions of Columbia University's Earth and Environmental Engineering Program; the Center for Earth Systems Engineering and Management at Arizona State University; Cornell University's Science of Earth Systems major; the University of New Hampshire's Institute for the Study of Earth, Oceans, and Space; and the Center for Integrated Earth System Science at the University of Texas, Austin. Resources for modeling Chesapeake Bay were developed from various peer-reviewed literature sources and information reported and compiled by the Chesapeake Bay Foundation, the U.S. Environmental Protection Agency, NASA, NOAA, the Chesapeake Bay Environmental Observatory, and the Chesapeake Community Modeling Program.

The author would like to acknowledge Professor Maria Papadakis for assisting with the development of the ideas and content of Chapter 7. The author is indebted to the work compiled by the Center for the New Energy Economy (CNEE) in partnership with The Nature Conservancy (TNC), who built the State Policy Opportunity Tracker (SPOT) for Clean Energy to serve as a hub of information on both existing state clean energy policies and, uniquely, future policy opportunities. The author appreciates the leaders of the U.S. Energy History Visualization project and Liz Moyer, University of Chicago Department of the Geophysical Sciences and RDCEP, for developing an interactive model to help students understand how past U.S. energy transitions can help us understand our potential future path. The research lead is Robert Suits, University of Chicago Department of History. Graphics were built by Nathan Matteson, DePaul University College of Computing and Digital Media, and members of the University of Chicago's Research Computation Center, including Ramesh Nair, Milson Munakami, Kalyan Reddy Reddivari, Sergio Elahi, and Prathyusharani Merla, with the assistance of Benjamin Kleeman, DePaul University. The author would also like to thank the developers of the En-ROADS energy and climate simulation at Climate Interactive, Ventana Systems, the UML Climate Change Initiative, and the MIT Sloan Business School for development of interactive energy and climate learning tools.

Dr. Striebig would like to acknowledge the support of his parents Janet and Ronald for making education a priority. Dr. Striebig is also indebted to his children, Echo and Zachary, who are a constant inspiration for the hope and potential of future generations. Last but certainly not least, the author wishes to thank the Global Engineering team at Cengage Learning for their dedication to this title: Timothy Anderson, Senior Portfolio Product Manager; MariCarmen Constable, Learning Designer; and Alexander Sham, Content Manager.

Bradley A. Striebig

About the Author

Bradley A. Striebig

Professor of Engineering, James Madison University, Harrisonburg, Virginia

Professor Striebig earned his PhD from Pennsylvania State University. He is a founding professor of the engineering program at James Madison University and previously taught engineering at Gonzaga University and Pennsylvania State University. He has served as editor on major journals in environmental engineering and sustainable development. He has led major, funded, award-winning research activities focused on working with developing communities and natural treatment systems. He has published two textbooks on sustainability and engineering and has authored over 100 technical publications, including several book chapters, numerous peer-reviewed journal articles, and peer-reviewed conference proceedings.

Contributing Authors

Daniel Castaneda

Assistant Professor of Engineering at James Madison University, Harrisonburg, Virginia

Daniel received his PhD in Civil Engineering from the University of Illinois Urbana-Champaign. His major research interests are in the sustainable use of infrastructure materials in the context of an increasingly diverse society.

Jason Forsyth

Associate Professor of Engineering at James Madison University, Harrisonburg, Virginia

Jason received his PhD in Computer Engineering from Virginia Tech. His major research interests are in wearable computing to develop safety systems that provide continuous monitoring and sensing to protect human life.

Shraddha Joshi

Assistant Professor of Engineering at James Madison University, Harrisonburg, Virginia

Shraddha received her PhD in Mechanical Engineering from Clemson University. Her major research interests are in engineering design, engineering education, and connected products and systems design.

Digital Resources

MindTap Reader

Available via our digital subscription service, Cengage Unlimited, **MindTap Reader** is Cengage's next-generation eTextbook for engineering students.

The MindTap Reader provides more than just text learning for the student. It offers a variety of tools to help our future engineers learn chapter concepts in a way that resonates with their workflow and learning styles.

- **Personalize their experience**

Within the MindTap Reader, students can highlight key concepts, add notes, and bookmark pages. These are collected in My Notes, ensuring they will have their own study guide when it comes time to study for exams.

2.1 Introduction

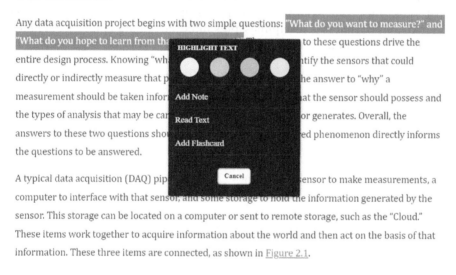

Any data acquisition project begins with two simple questions: "What do you want to measure?" and "What do you hope to learn from th[...]" [...] to these questions drive the entire design process. Knowing "wh[...] [...]tify the sensors that could directly or indirectly measure that p[...] [...]he answer to "why" a measurement should be taken infor[...] [...]at the sensor should possess and the types of analysis that may be ca[...] [...]or generates. Overall, the answers to these two questions sho[...] [...]ed phenomenon directly informs the questions to be answered.

A typical data acquisition (DAQ) pip[...] [...]sensor to make measurements, a computer to interface with that sensor, and some storage to hold the information generated by the sensor. This storage can be located on a computer or sent to remote storage, such as the "Cloud." These items work together to acquire information about the world and then act on the basis of that information. These three items are connected, as shown in <u>Figure 2.1</u>.

- **Flexibility at their fingertips**

Students can personalize their study experience by creating and collating their own custom flashcards. The ReadSpeaker feature reads text aloud to students, so they can learn on the go—wherever they are.

Cengage Read

Cengage Read, the new mobile app, enables students to study on the go. Students can read and listen to Cengage eTextbooks, and highlight and take notes when and where it's most convenient, both online and off. Best of all, it's free to download on any smartphone or tablet.

To learn more and download the mobile app, visit www.cengage.com/mobile-app/.

Cengage Unlimited

All-You-Can-Learn Access with Cengage Unlimited

Cengage Unlimited is the cost-saving student plan that includes access to our entire library of eTextbooks, online platforms and more—in one place, for one price. For just $124.99 for four months, a student gets online and offline access to Cengage course materials across disciplines, plus hundreds of student success and career readiness skill-building activities. To learn more, visit www.cengage.com/unlimited.

Engineering Concepts and Simple Tools: Materials, Mass, Gravity, and Moment Arms

Objectives

Chapter 1 introduces the basic principles of engineering analysis. The concepts of mass, weight, materials, and applied forces are investigated. Students will develop basic knowledge related to mass, weight, and forces acting on materials. They will create connections between material choices and material properties, and they will also gain an understanding of how various materials affect engineering design.

Student Learning Objectives

- Interpret engineering drawings to relate principles and equations in geometry.
- Create an engineering drawing to design and build a balance for two objects.
- Identify basic tools and safety procedures.
- Calculate the volume and density of various materials.
- Perform a force balance to balance a hinged load.
- Construct an engineered mechanism to balance a point and distributed forces.

This chapter discusses fundamental engineering design philosophies and properties that are addressed through the development of an integrated experiential-learning approach for engineering science, including applications of force and weight, force distribution, density, specific gravity, and applied geometry.

1.1   Introduction

If you recently started college, you might have had to answer a lot of questions recently, such as the following:

- What are you planning to do?
- Where are you going to college?
- What is your eye color?
- What are you going to major in?
- Do you drive a car?
- How much carbon dioxide (CO_2) did you emit today?

Most questions we are asked involve the **future**. Engineers are expected to be able to predict the future. They predict how long it will take to fly to Amsterdam or how long it will take to transport a load of goods via boat from Busan, Korea, to Los Angeles in the United States. How much can the boat or plane carry? How long will the water last that is stored in a reservoir? How much carbon dioxide is emitted from the cement required to build a nuclear power plant or a reservoir? If you plan on being an engineer, you should get used to reliably predicting the future!

If you are thinking about pursuing an engineering career, some questions you should consider to guide your education are the following:

1.1 **What are you currently doing, and why?**

1.2 **What would you like to accomplish before you retire?**

The word *engineer* is derived from the Latin word *ingeniator*, which translates to "ingenious, to devise in the sense of construction of craftsmanship." Another closely related word is *ingenuity*. Engineering ingenuity refers to the way in which we create and design, build relationships, interact with the surrounding environment, communicate with one another, and solve problems.

Engineers come from significantly different levels of math, chemistry, and physics backgrounds. In addition to *science, technology, engineering, and math (STEM)* courses, successful engineers must be able to communicate their ideas to others through drawings, illustrations, and writing. Engineers must be proficient in STEM-related subjects to practice the applications of fundamental math and engineering science to solve problems. However, other learned experiences from art classes, English classes, and shop classes, to name only a few, will help engineers use common tools and communicate complicated ideas to others. The communication piece involves oral and written skills to explain and promote ideas to stakeholders, investors, and the public. Good communication involves listening, understanding the cultural setting, accepting unbiased information, and exhibiting good judgment. Those skills, and the ability to illustrate ideas through technical drawings, are foundational and are required to communicate ideas for fabrication and manufacturing, as illustrated in Figure 1.1.

Not only is an appropriate engineering education required to be employed as an engineer by most employers, but also, and perhaps more important, your education is required to ethically practice engineering and protect the public health and safety. At the World Summit on Sustainability in Johannesburg, South Africa, in 2002, the role of education was declared "to emphasize a holistic, interdisciplinary approach to developing the knowledge and skills needed for a sustainable future as well as changes in values, behavior, and lifestyles. This requires us to make decisions and act in culturally appropriate and locally relevant ways to develop and evaluate alternative visions of a sustainable future and to fulfill these visions through working creatively with others."

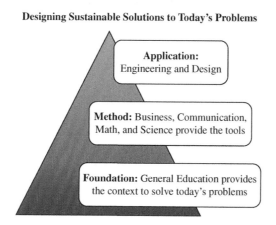

Designing Sustainable Solutions to Today's Problems

Application: Engineering and Design

Method: Business, Communication, Math, and Science provide the tools

Foundation: General Education provides the context to solve today's problems

Figure 1.1   Skills and the educational framework for engineering education.

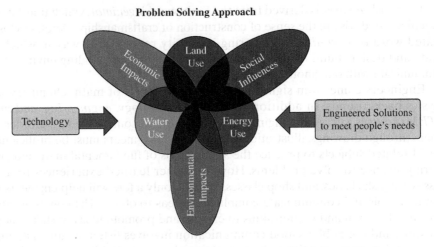

Figure 1.2 The engineering problem-solving approach is multifaceted and includes considerations of economic, environmental, and social impacts for sustainable design.

Engineers must consider the cost of their design from a business standpoint and end-user standpoint to justify decisions about energy, materials, transportation, and other costs associated with manufacturing goods. Engineers must also consider the availability and consumption of natural resources and the impacts of their designs on the environment, as illustrated in Figure 1.2. It is incumbent on engineers to perform their work ethically and consider the implications of existing and new technology on society as well.

1.3 Describe how the U.S. Bureau of Labor Statistics defines an engineer and an engineering technician (or engineering technologies) by reviewing the "field of degree" highlights for students and job seekers in the *Occupational Outlook Handbook*. Describe the difference between an engineer and a technician, and provide a salary range for each (www.bls.gov/ooh/field-of-degree).

1.4 The National Society of Professional Engineers (NSPE) has established a Code of Ethics for engineers. Look up and write down the six fundamental canons of the NSPE Code of Ethics.

1.2 Engineering Science and Practice

Engineers are often required to solve new and unfamiliar problems using their ingenuity. In this text and in your future classes, you will learn a systematic approach to solving engineering problems. This approach is common across engineering disciplines and involves breaking the problem down into as many small pieces as possible. Any solution to a problem is not very useful until and unless it can be communicated to others, including other engineers, manufacturers, machine operators, and so on. In addition, by following a common systematic approach, your work can be reviewed by others to ensure its accuracy. In some cases, your work may be reviewed to address future problems or to prevent lawsuits and liability claims against your work. The engineering presentation shown in Figure 1.3 is a very important part of communicating your engineering design. Furthermore, as an added bonus, this method really helps in getting a good grade and helping your peers learn from your work!

Although not every professional engineer works on "engineering paper" illustrated in Figure 1.3, most engineers work on similar paper or notebooks with a format that is very similar to the problem-solving process shown. In this process, engineering paper is used and recommended for all your engineering work. The top portion provides a space for labeling the work with your name, the assignment or project number for the project you are working on, the date the work is being done, and the page number.

The engineering problem-solving process involves five steps: problem definition, a sketch of the problem and possible solution, a list of known and unknown variables, a representation of the pertinent mathematical equations, and a detailed written mathematical analysis and solution. Each step is described below.

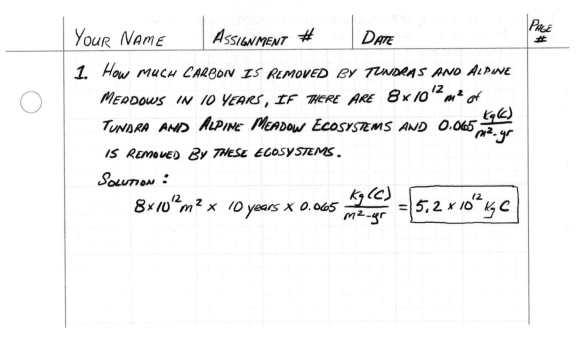

Figure 1.3   An example of the engineering problem-solving process expected for homework submissions. Note that the problem is stated and that the units associated with each given value are clearly shown.

1. *Problem definition*: The first step in solving any problem is to clearly articulate the problem and communicate the problem in writing. For homework problems, this usually means writing the problem or information that is given. For open-ended problems and professional work, this means writing down the portion of the problem you are trying to find or solve. Any assumptions made about the system should also be stated.

2. *Problem sketch*: Provide a sketch that represents and relates the problem elements. In some cases, this might include a map, a drawing of a reaction process, or a drawing of forces acting on a structure or system. The sketch should include as much numerical information that is known and identify elements that are unknown.

3. *Variable list*: Variables, both known and unknown, and constants that are relevant to the mathematical analysis should be listed. The system of units of each variable, which will be detailed in the next section, should be defined for each variable.

4. *Summary of mathematical equations*: All mathematical expressions and equations that relate to the problem should be identified and written down, using the variables, constants, and given information from the list created in step 3.

5. *Mathematical analysis and solution*: Solve the mathematical system for unknown variables, showing all algebraic substitutions and relevant mathematical processes. This should include inspection and analysis of the units used to ensure that the dimensions and units in the solution are consistent with those given in the problem definition in step 1. It is customary and often helpful to outline, or "box," the final solution so it can be quickly and easily identified.

1.3 Fundamental Unit Analysis

The application of science and engineering requires a proficiency with numbers and understanding measurements. Lord Kelvin expressed this in saying, "I often say that when you can measure what you are speaking about, and express it in numbers, you know something about it; but when you cannot measure it, when you cannot express it in numbers, your knowledge is of a meager and unsatisfactory kind: it may be the beginning of knowledge, but you have scarcely, in your thoughts, advanced to the stage of science, whatever the matter may be."

Measurements are used to describe a system and define standards. For measurements to be useful to an engineer, they must be reproducible and consistent. Measurements can be made of fundamental units as well as derived quantities. The units and measurements are common in our everyday lives. For example, a kilometer is a fundamental unit of length, and an hour is a fundamental unit of time. Your speedometer in your car measures the derived quantity, velocity, and the length, L, driven divided by the time, T, over which that length is driven to determine the velocity, or speed in meters per second (or miles per hour in the United States):

$$\text{Velocity} = \frac{\text{Length}}{\text{Time}} = \frac{L\,[\text{m}]}{T\,[\text{s}]}$$

Fundamental quantities are essential physical quantities that can be measured independently of one another. Fundamental quantities are listed in Table 1.1. Dimensions describe the basic concepts and physical nature of measured quantities. The dimensions have the same names as the fundamental quantities.

Derived quantities are related to and developed from the fundamental quantities. There are many, many derived quantities with different units. A few commonly used derived quantities include the following:

- Force
- Velocity
- Energy
- Volume
- Density
- Pressure
- Power

For example, the volume of a square is derived as the length expressed in units of meters (as shown in the square brackets) multiplied by the length of the height of the square and then multiplied by the length of the depth of the square:

$$\text{Volume, } V = L\,[\text{m}] \times L\,[\text{m}] \times L\,[\text{m}] = L^3\,[\text{m}^3]$$

Table 1.1 Fundamental quantities, dimensions, and SI units

Fundamental Quantity	Length	Mass	Time	Temperature	Electric Current
Dimension	L	M	T	θ	Q
SI Unit	Meter	Gram	Second	Kelvin	Ampere
SI Abbreviation	m	g	s	K	A

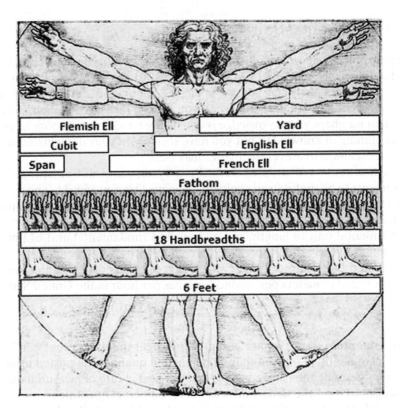

Figure 1.4 The common English units of the foot, the yard, or the hand were convenient to use in earlier times and were based on common human proportions.
Source: Unitfreak/Wikimedia Commons.

Every measured quantity will have both a numerical value and a unit of measurement. The units of measurement are the means of expressing the quantity and comparing it to a known *reference*. In earlier times, units were based on the proportions of a person as the reference, as illustrated in Figure 1.4.

The *American engineering system* is based on a standardized version of these old English measurements. The fundamental units of the American engineering system are the foot (ft) for length, the pound-mass (lb_m) for mass, and the second (s) for time, as shown in Table 1.2. For historical reasons, many of these units are still commonly utilized by engineers in the United States. However, the historical basis of these units makes it more difficult than the new international measurement system to use in most cases.

Table 1.2 Fundamental quantities and units in the American engineering system

Quantity	Unit	Symbol
Length	Foot	ft
Mass	Pound (mass)	lb_m
Time	Second	s
Temperature	Degrees Fahrenheit	°F
Molar amount	Pound mole	lb-mol

In 1960, an international conference formulated the "Système Internationale d'Unités," or the *SI* system, which has been widely adopted by the global scientific and engineering community. This system is based on physical phenomena that can be measured under specific conditions. The system uses the density of water to define the standard units of mass and volume; this makes relating units of mass and volume for many substances much easier than in the American system. The SI system also uses unit prefixes based on powers of 10, again making conversion between large and small measurements much simpler than the American system. The most common of these prefixes and their abbreviations for the SI units are shown in Table 1.3.

Engineers use the language of numbers, measurements, and units. Engineers practicing in the United States must be fluent in both the language of units used in the American engineering system and the SI system of measurements. Engineers are commonly required to convert measurements between these systems of units. It is acceptable to look to direct conversion factors in appropriate references or from reliable sources on the internet. Often, engineers might find themselves in the field when online resources and references are not available; at such a time, it is useful to use conversion factors that can be derived from fundamental dimensions using conversion factors you may remember from common use. However, be very careful when converting units, as mistakes related to the use and conversion of units have resulted in many engineering mistakes. For example, on September 23, 1999, NASA lost the $125 million Mars Climate Orbiter space probe after a 286-day journey to Mars. Miscalculations due to the use of AE units instead of SI units apparently sent the craft slowly off course—60 miles in all. On January 26, 2004, at Tokyo Disneyland's Space Mountain, an axle broke on a roller coaster train mid-ride, causing it to derail. The cause was a part being the wrong size due to a conversion of the master plans in 1995 from English units to metric units. In 2002, new axles were mistakenly ordered using the pre-1995 English specifications instead of the current metric specifications. When doing engineering calculations, include your units in your calculations and carefully and clearly write them down!

Table 1.3 Prefixes for SI units based on powers of 10

Power	Prefix	Abbreviation	Power	Prefix	Abbreviation
10^{-24}	yocto	y	10^{3}	kilo	k
10^{-21}	zepto	z	10^{6}	mega	M
10^{-18}	atto	a	10^{9}	giga	G
10^{-15}	femto	f	10^{12}	terra	T
10^{-12}	pico	p	10^{15}	peta	P
10^{-9}	nano	n	10^{18}	exa	E
10^{-6}	micro	μ	10^{21}	zetta	Z
10^{-3}	milli	m	10^{24}	yotta	Y
10^{-2}	centi	c			
10^{-1}	deci	d			

1.5 Describe the problems with using an ancient human proportion–based measurement system if you were to build a horse-drawn carriage and you were to purchase parts for the carriage and horses from 12 different suppliers. You are also concerned about navigating under a bridge that is one fathom high that you must pass under to get between your work and your home.

1.6 Force is a derived quantity determined from the acceleration of an object multiplied by the mass of an object. Derive the dimensions used to create units for a force acting on an object from the fundamental quantities of length, mass, and time. Also, show the SI units of force in your derivation.

1.7 The highest mountain in the world is Mount Everest in Nepal. The peak of Mount Everest is 29,029 feet above sea level. How many miles high is this? What is the elevation in meters?

1.4 Analysis of Forces

Scalar quantities have only one magnitude. Mass is a scalar quantity that describes how much matter is contained in an object. An example could be the mass of a soft drink container, typically about 500 grams. Notice this has one value, 500, and one unit, grams.

Distance is also a scalar quantity. An example is the distance from New York to Los Angeles by airplane, which is 3,944 km. From this distance, one could perhaps calculate how much fuel would be needed for a given aircraft to make that trip. However, this distance information is not enough information to tell someone *how* to get from New York to Los Angeles, even if that person were piloting the airplane.

In order to tell someone *how* to get to Los Angeles from New York, you also need to provide them a direction: Fly west by southwest. . . . A vector quantity requires a value of some magnitude, units, *and* a direction. For example, the plane's vector may be flying at a velocity of 900 kilometers per hour at 15 degrees south of due west. This would allow us to calculate an estimated time of arrival (ETA) for the flight. Notice also that the directional part of the vector must reference a coordinate system; in this example, the coordinates were in degrees, and the reference system consisted of points on a compass.

A coordinate system is used to specify locations in space. The coordinate system should have a fixed reference point that may be stated in the problem. The coordinate system also requires specified axes and instructions that label points in space relative to the origin of the axis. The most used coordinate system is the Cartesian coordinate system illustrated in Figure 1.5. A vector direction is the direction of the arrow given by an angle relative to the coordinate system, as illustrated in Figure 1.6.

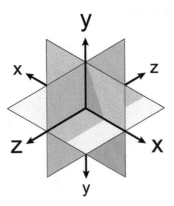

Figure 1.5 The Cartesian coordinate system.
Source: V_ctoria/Shutterstock.com.

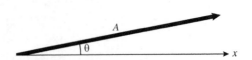

Figure 1.6 The direction of the vector *A* with an angle θ shown with respect to the *x*-axis.

1.5 Newton's Law

An object at rest tends to stay at rest unless a force acts on that object. Sir Isaac Newton observed this, as likely did many other members of humanity. However, Newton described this property mathematically and wrote down this mathematical observation as *Newton's first law of motion*. In addition, Newton noted that equilibrium implies that a body is either at rest or moving with constant velocity when the sum of the forces acting on that body are equal to zero:

$$\sum F = 0$$

An object will remain or move with uniform motion at a constant velocity until an external force acts on it.

When a force is imparted on an object, that force also must have a direction. A force has a magnitude and a direction and thus is expressed as a vector quantity. Newton observed and wrote down *Newton's second law of* motion that a force, F, is proportional to the acceleration, a, that the force produces on a given mass, m.

$$F = m \times a$$

A commonly described force is the weight of an object. The weight (W) of an object is the force acting toward the center of gravitational pull on the mass of an object multiplied by the acceleration of gravity. This weight will accelerate a parachuter toward the ground who jumps out of that airplane traveling from New York to Los Angeles. In this case, the parachuter is traveling and accelerating in a downward direction (prior to opening the parachute) since the force is also acting downward.

Calculating Mass and Density

Worker bees are very strong animals and can carry quite a bit of pollen back to the beehive. Several YouTube videos claim to show how much weight a bee can lift. Since you have developed a reputation as a luminary design engineer, your college bio-geneticist friend has asked you to help them create the world's first bee-based video drone using a GoPro video camera.

Your bio-geneticist friend conducted an experiment and collected the following data:

Number of beads a bee can lift	8	beads
Mass of a bead	12	milligrams
Mass of a GoPro	158	grams
Dimensions of the GoPro	71.8 $\times$ 50.8 $\times$ 33.6	millimeters
Number of bees that can fit on a square-inch surface	10	bees/in.^{2}

1.8 How many bees would it take to lift the GoPro for a BeeDrone Cam?

1.9 How many bees can fit on top of the GoPro to carry it?

1.10 How does the number of bees required compare to the number of bees that would fit on the top of a GoPro? Does your friend's idea have wings; that is, is the BeeDrone a good idea or not?

Can a Big Bug Create a Lethal Force? (Inspired by *Mythbusters*, Season 9, Episode 9)

Have you ever heard the expression *Sometimes you're the windshield, sometimes you are the bug?* (Sir Isaac Newton perhaps would have written a song by this title if he were a songwriter, but alas he was not. Feel free to look up the song by this title on your favorite music service.) How much damage can a little bug cause? If you are riding a bicycle or motorcycle at a high velocity and hit a flying bug, can you feel that? How hard did you hit that bug, which, in engineering terms, means, how much force was imparted onto you when you hit the bug? And what if that bug was a big bug; does size matter in this case? What if it hit you under the neck in a vulnerable area of your throat; could that cause serious life-threatening harm?

Let's answer these questions using the engineering problem-solving approach.

The first step is to write the specific questions we will solve in appropriate engineering language, which will help us prepare our mathematical analysis. Let's evaluate this question first: How hard did you hit that bug, which, in engineering terms, means, how much force was imparted onto you when you hit the bug? Specifically, this is the force imparted as the bug moves from its traveling velocity and decelerates to a "dead" stop when it hits your body.

Table 1.4 Deceleration of a theoretical "bug" on impact with an object

Bug Weight [g]	Bug Weight [kg]	Deceleration [m/s²]	Force [N]
0.1		450,000	
10		17,000	
60		7,500	

1.11 Determine the bug weight in kilograms and enter the data from Table 1.4 into a spreadsheet program like Excel or Google Sheets.

1.12 Calculate the force in newtons and plot the data from Table 1.4 in the spreadsheet.

1.13 Graph the data point in an Excel graph with acceleration (*y*-axis) vs. mass (*x*-axis).

   a) Plot the data on a scatterplot.
   b) Insert a power function trend line.
   c) Note the equation and R^2 value of the trend line on the graph.

1.14 Use the trend line equation to calculate the mass of a bug needed to impact exactly the lethal force required by a blow to the larynx (338 N).

Recall that, as engineers, we want to be able to predict the future. We do this by making observations and interpolations from those observations. We can create enough simulated bug force to seriously injure a person by firing a simulated bug out of a high-pressure air cannon, but could we experience that acceleration riding a bicycle or motorcycle? To determine this answer, let's examine the experimental date shown in Table 1.5.

Table 1.5   High-powered simulated bugs shot out of a cannon to impart lethal force

Motorcycle Velocity [m/s]	Bug Weight [kg]	Deceleration [m/s²]	Force [N]
40	0.06	(−) 7,500	
54	0.06	(−) 21,000	
90	0.06	(−) 40,000	

1.15 Determine the force from the impact of a bug in units of newtons for a motorcycle traveling at 40 m/s and a 60-g bug using the previous data to interpolate the required values. Write the relevant equations. Label your variables. Show your algebraic work and any unit conversations required. Calculate the force associated with each different velocity and complete Table 1.5 using a spreadsheet.

1.16 Graph the data point in an Excel graph with acceleration (*y*-axis) vs. velocity (*x*-axis).

a) Plot the data on a scatterplot.
b) Insert a logarithmic function trend line.
c) Note the equation and R^2 value of the trend line on the graph.

1.17 Estimate the force imparted when a 60-g bug impacts an object if the speed differential between the bug and the object is 45 m/s. Use the acceleration by interpolation or from the graphical data. Show your calculations.

1.18 Determine the equation that relates the force imparted to a motorcycle driver at various velocities from a curve fit of the graphical data. Use this relationship to determine the speed that a motorcycle would have to be traveling in order to impart a lethal force (338 N) to a motorcyclist from a 10-g cicada.

1.6   Visualization—Engineering Drawing

It is critical for engineers to be able to communicate their ideas precisely through the spoken word, written word, and appropriate illustrations. There are specific rules used to create and communicate with technical drawings just like there are rules we learn in speaking and writing. These rules help engineers consistently and accurately communicate their ideas. Mechanical engineers may need drawings to communicate how a part will be machined. Electrical engineers use drawings to communicate circuit schematics and circuit board layout. Civil engineers use "blueprints" (older large printers used blue ink, hence the term "blueprint") to show how to construct buildings, bridges, and roads. In this section, you will learn how those ideas are communicated, first by learning how to "read" technical drawings and then by creating technical drawings yourself!

You have already been introduced to the specific process engineers use to solve problems and document their mathematical analysis. Technical drawings also have a specific format shown in Figures 1.7 through 1.9. Notice that each drawing has a border and an information box. The information box contains important information, including the following:

- The drawing title and ID number
- The dimensions, units of measure, and scale of the drawing
- A place for identification of who drew the illustration, who checked the illustration, and who approved the drawing. (Typically, a drawing is approved or "stamped" by a licensed professional engineer.)

There are also several standard types of lines used in engineering and technical drawings:

Continuous lines represent visible edges and boundaries and are relatively thick lines of 0.5 to 0.6 mm.

Hidden lines represent edges and boundaries that are not visible from the perspective drawn and are illustrated by dashed lines that are typically 0.35 to 0.45 mm thick.

Center lines are used to represent axes of symmetry and consist of a long dash followed by a short dash and are typically thinner lines 0.3 mm thick.

Phantom lines are used to indicate alternative positions of moving parts or adjacent positions of related parts. These lines look similar to the center lines but have two short dashes, as shown in Table 1.6.

Dimension and extension lines are used to indicate the length of an object's feature. The feature's measurement, called the dimension, is the numerical value typically found in the center of two extension lines and a broken line terminated by two arrowheads as shown below.

Cutting plane lines show where an imaginary "cut" has been made so that a part of an object hidden by viewing only the outside of the object can be visualized and illustrated. Arrows are used at both ends of the cutting line to indicate the direction the object would be viewed if the imaginary cut was made.

Section lines are used to show the areas that have been cut by the cutting plane line. These lines are drawn over the area typically in a 45-degree pattern, as illustrated in Table 1.6, and are much thinner than continuous lines, typically only 0.3 mm thick.

Break lines are curved lines used to show imaginary breaks in a long object without any changes in the intervening length and are typically thicker lines of 0.75 mm. These lines are typically used to illustrate dimensions of long objects, like a two-inch by 12-foot section of lumber or a long pipe at a scale suitable to the size of the paper available.

Table 1.6 Line types used in engineering and technical drawings

Continuous line	———————
Hidden line	- - - - - - - - - -
Center line	—— — ——
Phantom line	—— — — ——
Dimension and extension lines	|←—— 10 ——→|
Cutting plane lines	↑ ⌐ — — — ⌐ ↑
Section line	/////////
Break lines	∿

Reading Engineering Drawings

The engineering drawing shown in Figure 1.7 is a metal counterbalance you will use as part of a balance scale. The drawing shows an aluminum part that is square on the top and bottom surfaces and has a rectangular depth. Notice in this drawing that both a two-dimensional and a three-dimensional representation are shown. "Read" this drawing and answer the following questions:

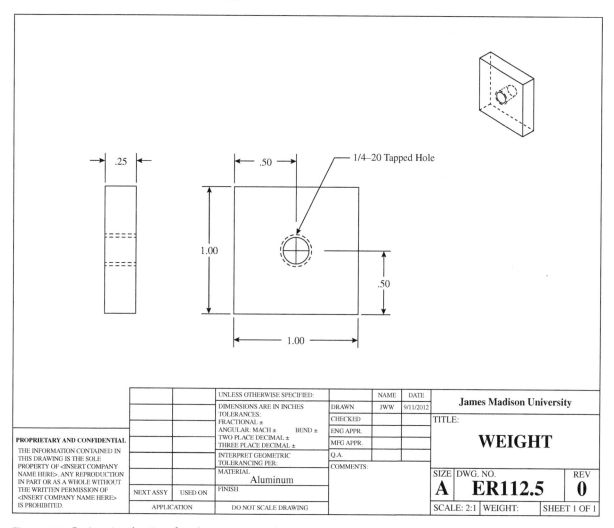

Figure 1.7 Engineering drawing of an aluminum counterbalance used to illustrate engineering drawing components and provide data for calculating the force acting on the end of the balance beam.

1.19 What are the initials of the person who made this drawing?

1.20 What are the units of measurement shown on the drawing?

1.21 What is the scale of the drawing?

1.22 What is the length of each side of the square top of this part?

1.23 What is the depth of the side of this part?

1.24 Label one example of the following line types on the drawing shown in Figure 1.7:

a) Solid line
b) Hidden line
c) Dimension line

1.25 Do the following for the aluminum counterbalance shown in Figure 1.7.

a) Record the dimensions of the part's height, width, and depth.

b) Calculate the material volume of the part. Remember that you must consider the hole in the part when calculating the volume of the part's material.

c) If the aluminum has a density of 2.7 g/cm³, what is the mass of the counter-balance part shown in Figure 1.7?

d) If the counterbalance was hung on the end of a nail, what is the force that would be applied on the end of the nail? In which direction does the force applied to the nail act?

Drawings can be shown in two-dimensional or three-dimensional projections. Both dimensional views have value and may be found on the same drawing page, as shown in Figure 1.7. In the case of this simple symmetrical part, two two-dimensional drawings are sufficient to illustrate everything needed to fabricate this object. For more complicated parts that are not symmetrical, it often requires three unique two-dimensional drawings to convey all of the information required to fabricate a three-dimensional part. Orthographic projection is the systematic method of portraying a three-dimensional object through two-dimensional drawings. Typically, this involves using your imagination to revolve the object in your mind so that you are looking directly at different sides of the object.

A small speaker wall mount is shown in Figure 1.8. There are six principal perspectives or views that can be projected by looking directly toward one of the six possible sides, as illustrated in Figure 1.8. Usually, technical drawings involve a perspective from directly above the object looking directly downward onto the top of the object, as shown in Figures 1.8 and 1.9, and this projection is called the ***top view***. The largest and most detailed point of view of an object is typically defined as the front of the object, or ***front view***. The front of the object is chosen to yield the fewest number of hidden lines and is the first projection chosen. A third view from the side of the object that visually conveys the most information, as shown in Figure 1.9, is called the ***side view***. The side view is drawn to the right of the front view for a ***right-side view*** or to the left of the front view of an object for a ***left-side view***. For simple objects like the aluminum counterbalance shown in Figure 1.7, only two views may be required. However, for more complicated objects, four views (top, front, left-hand side, and right-hand side) may be necessary. All six views that are illustrated in Figure 1.8 can be utilized if each view is required to convey the size, scale, placement of cuts and holes, and any other information needed to fabricate or manufacture a part or object.

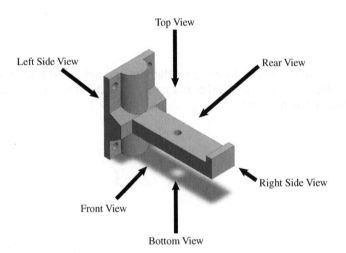

Figure 1.8   There are six possible viewpoints, or projections, for creating a two-dimensional drawing of a three-dimensional object.
Source: Jasmine White.

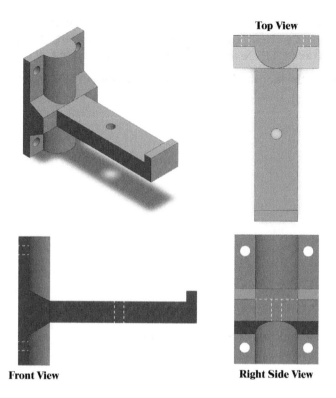

Figure 1.9 An example of orthographic projections of the top view, front view, and right-side view of a three-dimensional object. Hidden lines are shown in white.
Source: Jasmine White.

Designing a Simple Machine

Prior to the use of modern currency, payments were made by trading goods. Those goods could sometimes be paid for with precious metals: copper, silver, or gold. The copper, silver, and gold coins were literally "worth their weight," so a method had to be devised to determine the weight of the metal offered in copper, silver, and gold compared to the good being purchased. Merchants maintained scales or balances for this purpose. We still refer to this practice today when we "balance" our checking account by checking income versus expenses!

A balance is a simple mechanical device that illustrates many fundamental principles and can be easily built and assembled. By designing a balance according to the following procedures, you will be doing the following:

- Utilizing the density and specific gravity of solid and liquid materials to perform a force balance
- Utilizing the principles of geometry and algebra to calculate the weight and force acting on the balance beam
- Reading engineering drawings to relate numerical values to real parts and components of the balance.

1.26 Do the following given that water has a density of approximately 1,000 kg/m³ and that the PVC from which the pipe and endcap on the attached drawings (Figures 1.10 and 1.11) are made has a density of approximately 1,330 kg/m³.

a) Calculate the length of pipe, x, required to contain 56.7 g of water.

b) Calculate the mass of the pipe in grams.

c) Calculate the mass of the cap in grams.

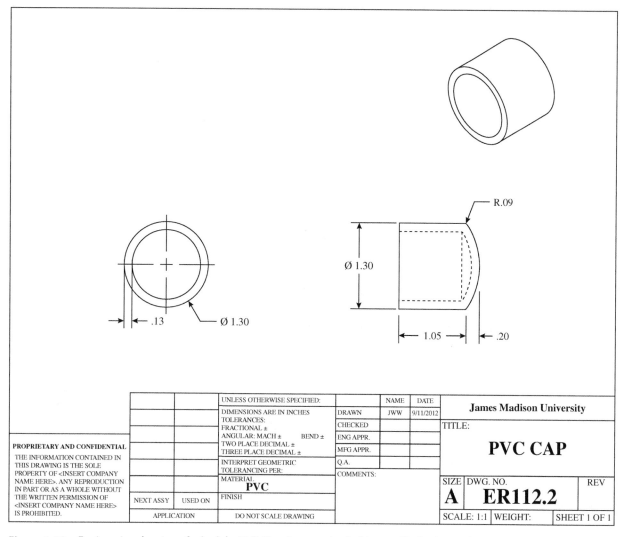

UNLESS OTHERWISE SPECIFIED:			NAME	DATE	**James Madison University**		
DIMENSIONS ARE IN INCHES		DRAWN	JWW	9/11/2012			
TOLERANCES: FRACTIONAL ±		CHECKED			TITLE:		
ANGULAR: MACH ± BEND ±		ENG APPR.					
TWO PLACE DECIMAL ± THREE PLACE DECIMAL ±		MFG APPR.			**PVC CAP**		
INTERPRET GEOMETRIC TOLERANCING PER:		Q.A.					
MATERIAL **PVC**		COMMENTS:			SIZE	DWG. NO.	REV
FINISH					**A**	**ER112.2**	
DO NOT SCALE DRAWING					SCALE: 1:1	WEIGHT:	SHEET 1 OF 1

NEXT ASSY   USED ON

APPLICATION

Figure 1.10   Engineering drawing of schedule 40 PVC endcap used to hold a specified volume of water in the balance.

d) Calculate the mass of the complete assembly of the pipe, water, and two caps.

Once you have completed this work, be sure to save the calculations and results, as you may use this information to fabricate the balance.

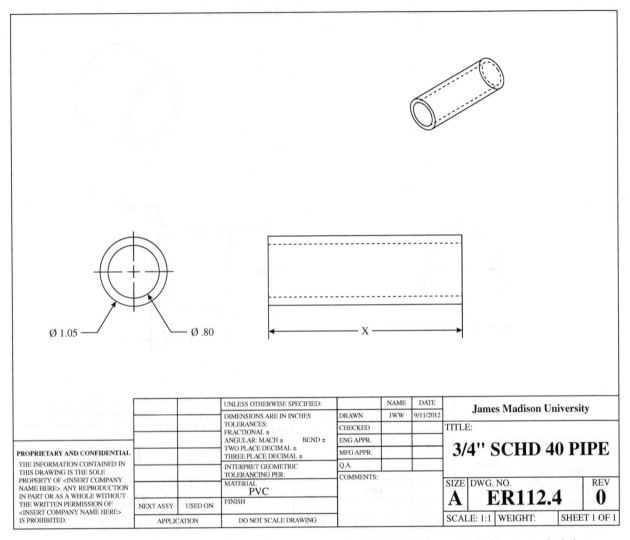

		UNLESS OTHERWISE SPECIFIED:		NAME	DATE	**James Madison University**	
		DIMENSIONS ARE IN INCHES	DRAWN	JWW	9/11/2012		
		TOLERANCES: FRACTIONAL ±	CHECKED			TITLE:	
		ANGULAR: MACH ± BEND ±	ENG APPR.				
		TWO PLACE DECIMAL ± THREE PLACE DECIMAL ±	MFG APPR.			**3/4" SCHD 40 PIPE**	
		INTERPRET GEOMETRIC TOLERANCING PER:	Q.A.				
		MATERIAL PVC	COMMENTS:			SIZE DWG. NO. REV	
						A **ER112.4** **0**	
NEXT ASSY	USED ON	FINISH					
APPLICATION		DO NOT SCALE DRAWING				SCALE: 1:1 WEIGHT: SHEET 1 OF 1	

PROPRIETARY AND CONFIDENTIAL

THE INFORMATION CONTAINED IN THIS DRAWING IS THE SOLE PROPERTY OF <INSERT COMPANY NAME HERE>. ANY REPRODUCTION IN PART OR AS A WHOLE WITHOUT THE WRITTEN PERMISSION OF <INSERT COMPANY NAME HERE> IS PROHIBITED.

Figure 1.11 Engineering drawing of schedule 40 PVC pipe used to hold water in the balance experiment.

1.27 Considering the precision that we will be able to attain in building the balance, is it acceptable to disregard the dome-shaped end of the cap and assume a flat shape for your calculations? Why or why not?

The piece of wood shown in Figure 1.12 is 1.75 cm × 3.6 cm × 20 cm and weighs approximately 64.5 g. Calculate the length of a wooden beam that is 50 cm + the last digit of your student ID number. For example, Amy's student ID number is 100055446. The length of Amy's beam will be 50 cm + 6 cm = 56 cm since her ID number ends with the number 6.

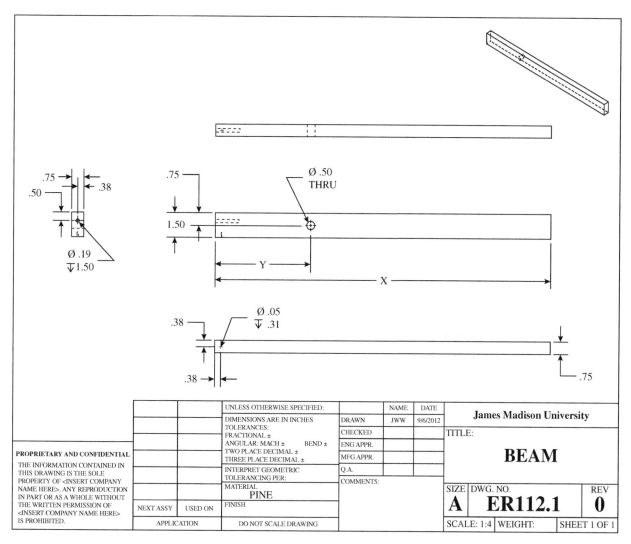

Figure 1.12 Engineering drawing of wood balance beam used to illustrate engineering drawing components and provide data to calculate the centroid of the weight-loaded beam.

1.28 Calculate the weight of a wooden beam that is 1.75 cm × 3.6 cm in cross section whose length, *X*, is 50 cm + *Z* cm, where *Z* = the last digit of your student ID number.

1.29 Add 47.5 g for the hardware and the mass of the wood to the weight of the aluminum counterbalance assembly.

1.30 The balance's beam shown in Figure 1.12 is to be held by a pin, placed at point Y. The aluminum bar will be held on a nail at one end. The PVC tube holding two ounces of water will be suspended from a nail at the other end of the beam. Sketch the beam and illustrate the point where the beam is attached to a vertical frame. Also show the forces and the direction of the forces acting on the beam from the aluminum counterbalance on the right-hand side of the beam and the water-filled PVC tube on the left-hand side of the beam.

The forces acting on the beam can be balanced by creating equal and opposite *moments* on each end of the beam. The ***moment*** of a force is a measure of the tendency of a force to cause a body to rotate about a specific point—in this case the pin in the balance. The equation for the moment of a force is

$$\text{Moment} = \text{Force} \times \text{Distance}$$

In order to balance the moment created by the aluminum counterbalance and the water weight, both moments must be equal to one another:

$$\text{Moment}_{\text{Aluminum}} = \text{Moment}_{\text{Water}}$$

1.31 For a beam of a given length, *X*, where *X* is the beam length determined in Problem 1.28, determine the position, *Y*, where the moments associated with the water-filled PVC tube will be equal to the moment associated with the aluminum counterbalance and the beam will be perfectly balanced horizontally. Recall that the force associated with each part was determined above.

1.32 Using tools provided to you, construct a balance and place it on a pin on a stand or one available in the classroom. Is your beam perfectly balanced (i.e., does it hang from the pin in a perfectly level and horizontal position)? Describe possible sources of error in the construction of the beam that might cause the beam to be less than perfectly balanced.

1.33 In the days before common currency, merchants may have developed poor reputations if people felt they were being cheated. Would it be relatively simple or difficult to change the way a scale behaved if a merchant were trying to balance coppers (copper coins) and receive an adjusted weight in silver? Describe how a dishonest merchant might "tip the scales."

References

Striebig, B., Ogundipe, A., and Morton, S. (2014). "Lessons in Implementing Sustainability Courses into the Engineering Curriculum." 121st ASEE Annual Conference & Exposition, Indianapolis, IN, June 15–18, 2014.

Striebig, B., Prins, R., and Wild, J. (2019). "Combining Basic Tool Training and an Introduction to Physical Sciences for Freshmen Engineering Students." ASEE First Year Engineering Experience, 11th Annual Conference, State College, PA, July 28–30, 2019.

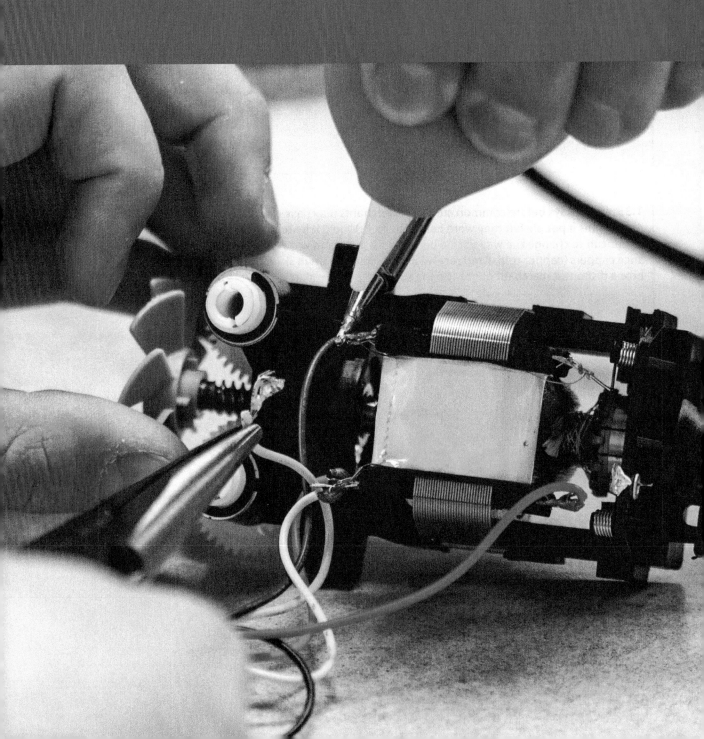

Objectives

This chapter discusses how engineers can assemble sensors and computers to collect data about the physical world. While the examples and components used will be from smaller hobbyist systems, the principles discussed apply to many more extensive systems.

Student Learning Objectives

- Describe the differences between "analog" and "digital" systems.
- Describe how engineers collect and use data from the natural world to inform decisions.
- Describe how information is transformed from electrical energy into binary information and then shared/sent through a communications channel.

2.1 Introduction

Any data acquisition project begins with two simple questions: "What do you want to measure?" and "What do you hope to learn from that measurement?" The answers to these questions drive the entire design process. Knowing "what" is to be measured helps identify the sensors that could directly or indirectly measure that phenomenon. More important, the answer to "why" a measurement should be taken informs the accuracy or reliability that the sensor should possess and the types of analysis that may be carried out from the data the sensor generates. Overall, the answers to these two questions should match such that the measured phenomenon directly informs the questions to be answered.

A typical data acquisition (DAQ) pipeline has three components: a sensor to make measurements, a computer to interface with that sensor, and some storage to hold the information generated by the sensor. This storage can be located on a computer or sent to remote storage, such as the "Cloud." These items work together to acquire information about the world and then act on the basis of that information. These three items are connected, as shown in Figure 2.1.

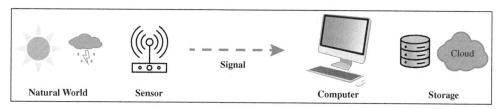

Figure 2.1 General data acquisition (DAQ) pipeline.

Throughout this chapter, we examine the decisions that are required for each "stage" of the pipeline. When selecting a sensor, computer, or storage medium, what are the important things one should consider in those phases? Overall, the design must work together to achieve the overall mission of sensing the phenomenon of interest.

2.1 What are some things of interest to you that you would like to measure? What would you hope to learn about these measurements?

2.2 Layers of Engineering and Science

Many people and disciplines will be involved when designing a data acquisition system. Problems that occur at the intersection of engineering disciplines create exciting opportunities for innovation. For the data acquisition system shown in Figure 2.1, the primary disciplines that might be involved are electrical engineering, computer engineering, and computer science. While these are vast fields, there is often a lot of historical and practical overlap.

To help better understand what an electrical engineering, computer engineering, or computer science person would care about, consider a typical smartphone, as shown in Figure 2.2. Many are familiar with the various applications, operating systems, and interfaces present when we utilize a smartphone. The design of these applications, ensuring that they run quickly and efficiently, and the management of the overall system are concerns of computer science. From a computer engineering perspective, the primary challenge is to select and integrate the various processors, sensors, and radio equipment that provide the smartphone's functionality. This separation of concerns is not a strict barrier. Many computer scientists and engineers will collaborate to develop the software and select the best hardware for the application. Furthermore, when examining the curriculum that someone studies in computer science or computer engineering, there is significant overlap, and it can

Application:
Facebook, Tik Tok,
Mail, Safari, Settings,
OSX

Hardware:
processor, graphics,
Wi-Fi, Bluetooth,
accelerometer,
gyroscope

Electrical:
battery, radio, buttons,
filters, speaker,
microphone, signals

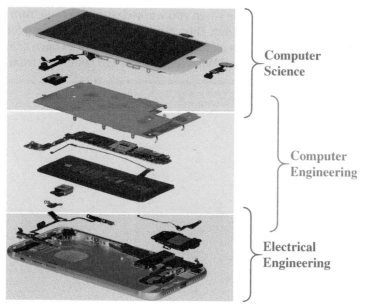

Computer
Science

Computer
Engineering

Electrical
Engineering

Figure 2.2 Hierarchy of electrical engineering, computer engineering, and computer science as seen through a smartphone.
Source: Adapted from IHS Markit Apple iPhone 7 teardown-exploded view (Photo: Business Wire) www.businesswire.com/news/home/20160920006782/en/iPhone-7-Materials-Costs-Higher-Than -Previous-Versions-IHS-Markit-Teardown-Reveals.

be easy for someone to move between the disciplines. If a distinction is made, it could be that computer science is "software focused" and a computer engineer is more "hardware focused," but each will write code and work with hardware in their daily tasks.

Underneath both layers is the electrical layer that powers the hardware, receives/sends communications, and physically designs the printed circuit boards that the entire smartphone requires. This area is most likely the responsibility of an electrical engineer. Like the relationship between computer engineering and computer science, there is much overlap between computer engineering and electrical engineering responsibilities. Computer engineers and electrical engineers work together to ensure that the hardware selected can be appropriately powered by the system and that the correct communications hardware is chosen for the desired speeds. Electrical engineering is one of the oldest engineering fields, and before computer engineering existed (in the mid-1980s), many electrical engineers had a computer science background so they could work as pre-computer engineers!

Overall, when designing a system with sensors, computers, networks, and analysis, many disciplines will be involved. Despite what a traditional academic catalog may indicate, in practice, the boundaries between these disciplines are loose; it is common for individual engineers and computer scientists to move around in various roles.

2.2 Consider a time when you worked with someone from a different background or interest. This could have been something for school or in your personal life. In that experience, where did you all have commonality? Where did you have differences? What lessons did you take away from that experience?

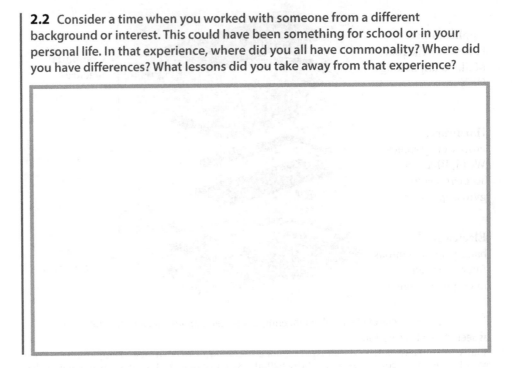

2.3 Selecting Sensors to Capture Physical Phenomenon

Searching for a suitable sensor can be overwhelming. There are many variations of each device, each with different sizes, costs, and robustness for surviving in the physical environment. This section covers the core parameters that need to be defined when selecting a sensor.

First, the selection process must begin with the question asked in Section 2.1: "What do you want to measure?" The answer to this question may be simple things, such as temperature, speed, or humidity, or it could be more complex, such as density, solar intensity, or air toxicity. There are often dedicated sensors for capturing that information for common or essential measurements. For example, a thermometer measures temperature, a barometer measures air pressure, and an accelerometer measures movement and the force of gravity. When a sensor directly measures the phenomenon of interest, it is called a "direct" measurement. However, if the sensor does not immediately capture the information needed but the information can be inferred from its measurement, it is called an "indirect" measurement. For example, an accelerometer attached to a car cannot report its speed, but that velocity can be inferred by watching the changes in acceleration over time. An example of some common sensors with their direct and indirect measurements is shown in Table 2.1.

Often, indirect measurements are very powerful and can be combined with several sources to identify complex scenarios. Recently, smartwatches have been shown to detect when someone is becoming sick. This is not because the smartwatch measures a person's health directly but because it can combine several indirect measures, such as heart rate, activity level, and sleep patterns; it can identify that someone is outside their everyday routine or has a lower energy level (Miller 2020). This decline is highly correlated but not necessarily a predictor of illness.

Table 2.1   Common sensors with direct and indirect measurements

Sensor Name	Direct Measurements	Indirect Measurements
Accelerometer	Acceleration (m/s^2)	Velocity, activity recognition, energy expenditure
Thermometer	Temperature (°C)	Health, device activity
Altimeter	Elevation (m) or air pressure (psi)	Location in a building, future weather trends
Ultrasonic range finder	Distance (m)	Velocity, occupancy in a room

2.3 From the list of sensors in Table 2.1, which would you use to measure how active a person is throughout the day? How would your choice of sensor change if it was worn on a person or if it was fixed in the environment?

2.4 Find a sensor that you believe is interesting on SparkFun (www.sparkfun.com). What kinds of interesting applications could the sensor be used for? How could they be helpful in daily life?

Once a sensor type (accelerometer, range finder, and so on) has been identified, the next challenge is to determine the operational parameters required for that sensor. There is not one standard "type" for any given sensor. Various manufacturers offer models with different costs, performance ranges, and operational environments. An analogy can be drawn to automobiles, where there are many options to choose from, and the challenge is to identify which is suitable for the task one wishes to accomplish. Both a sports car and a truck are vehicles that can help move cargo around, but only one would be suitable for carrying heaving loads to a construction site.

We identify three crucial parameters when selecting a sensor: *range*, *resolution*, and *sample rate*.

A sensor's measurement range indicates the "highest" and "lowest" values the sensor can reliably report. The phenomenon that is to be measured must fall within this range, or the sensor will fail to report a value or even operate. For example, consider a situation where a new engineer is selecting a thermometer to monitor the temperature within an industrial refrigerator. A typical range for that environment may be from −10°C to 10°C, so the selected sensor should be able to measure temperatures effectively within that range. Thus, a sensor with a range of −20°C to 20°C would be suitable, as it completely covers the expected range of the environment. However, it is important to note that a sensor may perform well in one context but not in another. A sensor that operates from −20°C to 20°C would work well for refrigeration but would not work well in an oven, which could expect higher temperatures above 200°C. Furthermore, a device designed to measure in a "lower" range may not be able to survive in this "higher" environment!

2.5 What would be the expected range for a device measuring human body temperature? What would be the range for measuring the temperature within a residential building?

Another consideration when selecting sensors is the device's resolution or accuracy. This value determines how precisely a sensor can measure a particular value. When reading about a sensor, this value might be reported as a plus-or-minus ($\pm$) measurement to indicate its precision. For the example of a thermometer, a device may report a precision of ± 1°C, implying that the device is accurate down to 1°C for measurements. This value indicates how much "error" the sensor may have in making a measurement. For a thermometer with precision ± 1°C, if the "true" value of some environment was 10°C, then the sensor may report 10°C ± 1°C, which means that 9°C, 10°C, and 11°C are all "correct" measurements for the same environment.

All sensors have error in their measurement; it is unavoidable, but the question for the engineer is how much error/accuracy is required for the given application. Engineers must select sensors that can provide equal or greater precision than what is required for a specific application.

2.6 If a device measuring distance has a precision of 5 m, what would be the range of possible measurement if an object was actually 10 m away?

2.7 Imagine you are selecting sensors for the next generation of autonomous vehicles. Each vehicle will need to know how close it is to other objects. What level of precision is needed in the distance sensors placed in the car? Justify your response.

The final parameter to consider is a sensor's sample rating. This value indicates how frequently a sensor can provide a new measurement. Often this value is described in terms of "samples per second" or Hertz (Hz). For example, if a sensor has a sampling rate of 20 Hz, that would imply that it can take 20 samples every second.

Much like range and accuracy, the sampling rate for the selected sensor must be appropriate for the given application. It should generally match (or exceed) the rate at which the phenomenon of interest changes. Returning to the weather station example, if it is desirable to know the temperature throughout the day, then the sensor should have a sampling rate of at least one sample per minute, or one sample per hour. Sampling at a lower rate (once per day) would not provide much information about the temperature throughout the day. However, sampling at a much higher rate (several times per second) may be too frequent, as temperatures in an outdoor environment do not change that rapidly.

Some sensors are "analog" or "continuous," meaning they can constantly provide new measurements or information. For these devices, a sampling rate constraint is still imposed on the computer system that is interfacing with the sensor. While the sensor may be able to change constantly, the computer will have its sample rate limits. The guidelines for ensuring that the sampling rate matches the phenomenon still apply in this situation.

2.8 What would be an appropriate sampling for monitoring a person's body temperature during exercise? Justify your response.

The parameters of range, accuracy, and sampling rate are important to know about a sensor. However, the question arises: where can that information be found? Any reputable sensor manufacturer will provide a data sheet that clearly spells out these performance characteristics (and so much more). Figure 2.3 shows a data sheet for an ultrasonic range finder that measures the distance to various objects. The sensor operates by sending out pulses of high-frequency sound (42 kHz) and measuring the time it takes for the pulse to rebound or return to the sensor. While there is a multitude of information on the data sheet, the three parameters discussed can still be found and are highlighted in the figure. The range and accuracy of the sensor are indicated in the text at the top of the document stating that the sensor can provide "range information from 6-inches out to 254-inches with 1-inch resolution." It is also noted that any object closer than six inches is reported as six inches. The sampling rate for the sensor is shown farther down as occurring every 50 ms or at 20 Hz. The conversion between the sample interval and Hertz is shown below:

$$\text{Sampling Rate} = \frac{\text{Samples}}{\text{per second}} = \frac{1}{0.050 \text{ s}} = 20 \text{ Hz}$$

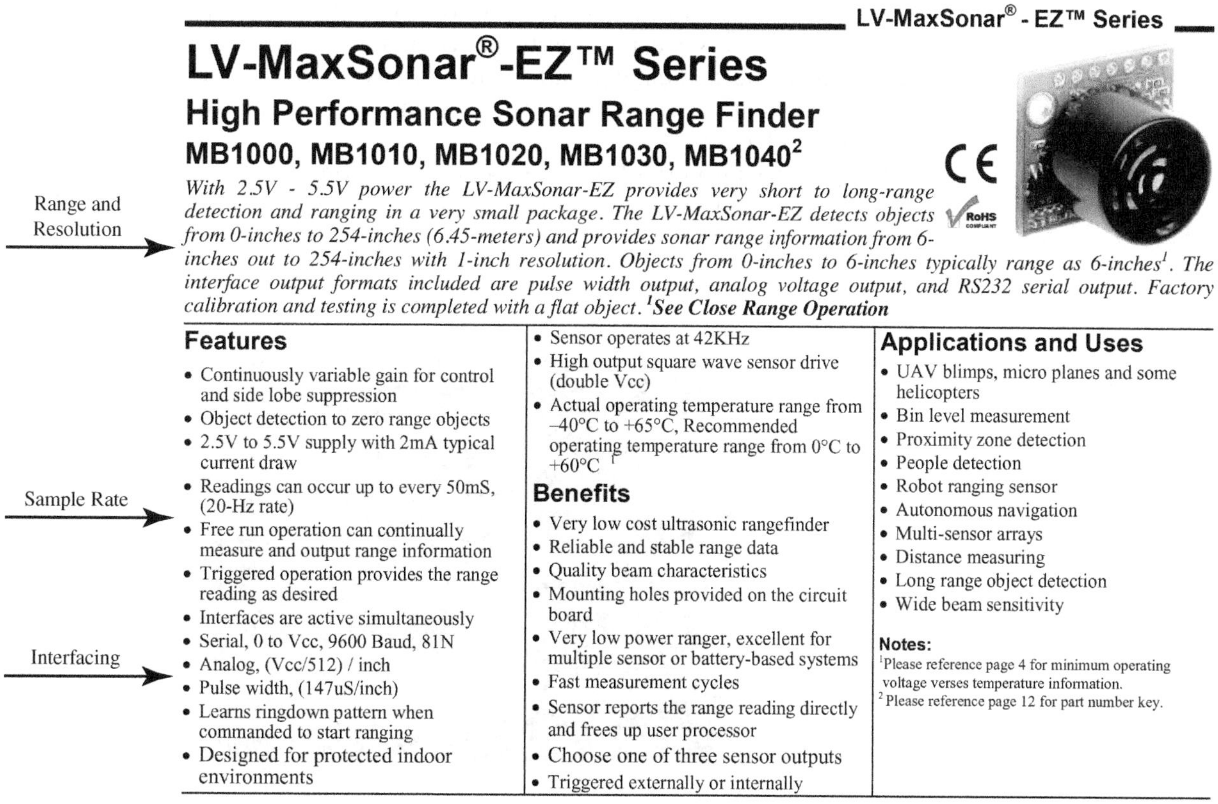

LV-MaxSonar®-EZ™ Series
High Performance Sonar Range Finder
MB1000, MB1010, MB1020, MB1030, MB1040[2]

CE
RoHS COMPLIANT

Range and Resolution →

With 2.5V - 5.5V power the LV-MaxSonar-EZ provides very short to long-range detection and ranging in a very small package. The LV-MaxSonar-EZ detects objects from 0-inches to 254-inches (6.45-meters) and provides sonar range information from 6-inches out to 254-inches with 1-inch resolution. Objects from 0-inches to 6-inches typically range as 6-inches[1]. The interface output formats included are pulse width output, analog voltage output, and RS232 serial output. Factory calibration and testing is completed with a flat object. [1]**See Close Range Operation**

Features

- Continuously variable gain for control and side lobe suppression
- Object detection to zero range objects
- 2.5V to 5.5V supply with 2mA typical current draw
- Readings can occur up to every 50mS, (20-Hz rate) ← Sample Rate
- Free run operation can continually measure and output range information
- Triggered operation provides the range reading as desired
- Interfaces are active simultaneously
- Serial, 0 to Vcc, 9600 Baud, 81N ← Interfacing
- Analog, (Vcc/512) / inch
- Pulse width, (147uS/inch)
- Learns ringdown pattern when commanded to start ranging
- Designed for protected indoor environments

- Sensor operates at 42KHz
- High output square wave sensor drive (double Vcc)
- Actual operating temperature range from –40°C to +65°C, Recommended operating temperature range from 0°C to +60°C

Benefits

- Very low cost ultrasonic rangefinder
- Reliable and stable range data
- Quality beam characteristics
- Mounting holes provided on the circuit board
- Very low power ranger, excellent for multiple sensor or battery-based systems
- Fast measurement cycles
- Sensor reports the range reading directly and frees up user processor
- Choose one of three sensor outputs
- Triggered externally or internally

Applications and Uses

- UAV blimps, micro planes and some helicopters
- Bin level measurement
- Proximity zone detection
- People detection
- Robot ranging sensor
- Autonomous navigation
- Multi-sensor arrays
- Distance measuring
- Long range object detection
- Wide beam sensitivity

Notes:
[1]Please reference page 4 for minimum operating voltage verses temperature information.
[2] Please reference page 12 for part number key.

Figure 2.3 Data sheet for EZ-1 ultrasonic range finder.
Source: © Maxbotix.

Another important element in the data sheet is a listing of the interfaces that the sensor provides. The trade-offs between various interfaces are discussed in the next section.

2.9 If a sensor completes a measurement once every five seconds, what is the sampling rate in Hertz? What if the device takes a measurement once every millisecond?

2.4 Acquiring Data from Sensors

Having selected a sensor, the next stage of the DAQ pipeline (Figure 2.1) is interfacing that sensor to a small microcontroller to collect and store the information. Like selecting sensors, picking a particular computer or microcontroller for a given application can be challenging, and a whole discussion of that topic is beyond our scope. The current discussion focuses on a common microcontroller, the Arduino Uno, shown in Figure 2.4.

The primary purpose of purchasing any microcontroller is its ability to interface with many sensors through its general-purpose input/output pins (GPIO). For the board in Figure 2.4, all the pins for the processor (the large rectangular

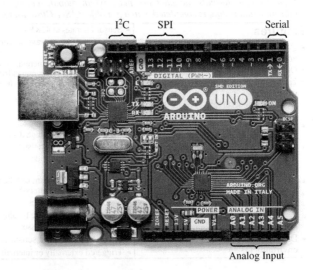

Figure 2.4   Arduino Uno with communications interfaces indicated.
Source: © SparkFun Electronics.

chip) have been broken out in each of the black headers that ring the perimeter of the board. This board has three general classes of pins: power, analog, and digital. The power pins provide electrical energy to turn on external sensors, and the analog/digital pins provide communication interfaces to collect information from those sensors.

The interface choice is important, as it must match the selected sensors. If it does not, the information will not be able to be transferred from the sensor to the computer.

Before discussing the analog and digital interfaces in more detail, a brief description of what is meant by "analog" and "digital" is required. From an engineering perspective, an analog signal is continuous and can take on any value between some upper and lower bound. If an analog signal is known to have a range between 1 and 10, it may take on any value between those bounds. The values 1, 2, 3.14159, and 7.13 are acceptable, as they fall between the upper and lower bounds. An example of an analog signal is shown in Figure 2.5 as a simple sign wave oscillating between 0 and 4.

Digital signals are different. These signals do not take on continuous values and exist only at a fixed number of discrete values. An example of a digital signal is shown in Figure 2.6 as a square wave. For this signal, only two values are possible: 0 when the signal is "LOW" and 2 when the signal is "HIGH." These HIGH and LOW measurements give rise to the binary 0s and 1s that make up the digital information in modern computing systems. In general, a LOW signal is considered binary 0 and would have a voltage at 0 V, whereas a HIGH signal is a binary 1 and would have a voltage near the system's operating level. These HIGH voltages are typically at 3.3 V or 5 V, depending on the architecture.

Analog and digital signals each have their advantages and disadvantages. An analog can enable more precise measurements due to its continuous values; however, noise can also be introduced into the analog signal from other components in the system that may corrupt the information. Digital signals are generally more

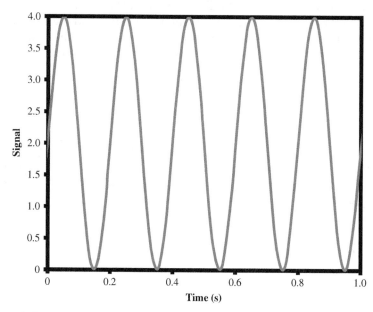

Figure 2.5   A sine wave.

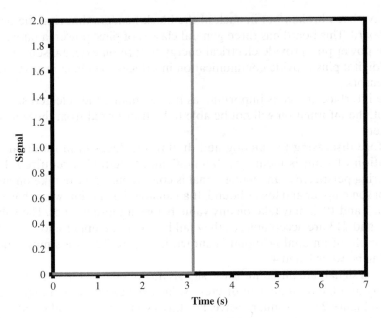

Figure 2.6   A square wave.

robust to noise, as the information exists only at distinct values; however, a digital interface may be more complex, depending on the number of wires in the interface.

Complicating the matter further is that various sensors can be produced with either analog or digital interfaces. Figure 2.7 shows a thermistor, an analog device that changes its properties based on temperature, and Figure 2.8 shows a custom environmental board that measures temperature and gas concentration. Each can accomplish the same tasks but may be more or less valuable, depending on the desired sensor characteristics (discussed in Section 2.3).

Several digital interfaces are available on the Arduino, the most notable being the Serial/UART port, Serial Peripheral Interface (SPI), and the Inter-Integrated Circuit (I2C). Each of these interfaces utilizes various pins on the Arduino to communicate with external sensors. In general, multiple pins are needed to send and

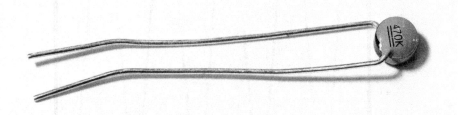

Figure 2.7   An "analog" leaded thermistor.

Figure 2.8   A "digital" BME688 environmental sensor.
Source: © SparkFun Electronics.

receive data from the Arduino and control that information's flow. Table 2.2 shows some of the common pin names for these roles with each interface and the typical speeds that can be achieved with each interface.

Examining these different pins, it is possible to understand some of the trade-offs between these other interfaces. The UART interface is straightforward, requiring only two pins, but it has no flow control and requires two communicating devices to operate closely in synch. The SPI interface has many pins that allow data to flow in/out of the Arduino simultaneously and with flow control. This can allow rapid exchange of information, but the interface is limited to only one or two devices. Finally, I2C is a good compromise between these systems. It has few connections, making it easy to wire up, and multiple devices can be placed on that interface. However, the trade-off is that it is much slower than SPI, and the transfer rate is shared among the various devices on the bus.

Given the various interfaces, many sensors provide multiple types to enable them to communicate with any microprocessor. The range finder in Figure 2.3 provides both a digital (Serial) and an analog interface. Although cryptic, the data sheet indicates that the Serial port is "9600 Baud, 81N," which indicates a transfer rate of 9,600 bits per second, with each message being eight bits long.

Table 2.2   Pins for different digital interfaces

Role	UART	SPI	I2C
Data in (to Arduino)	RX (pin 0)	CIPO (pin 12)	SDA (pin 18)
Data out (from Arduino)	TX (pin 1)	COPI (pin 11)	SDA (pin 18)
Flow control	None	SCK (pin 13) and SS (pin 10)	SCL (pin 19)
Typical transfer rate (bits per second)	115 Kbps	5 Mbps	1 Mbps

2.10 In selecting digital interfaces, which should be chosen to transfer the most data in the shortest possible time? If controlling the flow of data during transmission is important, which interface should *not* be selected?

While digital interfaces can have multiple wires for various purposes, an analog interface is a single interface that measures a voltage. The idea is that an analog or continuous sensor produces that voltage, which is then measured by the microcontroller. These analog interfaces are implemented with a unique device inside the microcontroller called an analog-to-digital converter (ADC), which measures the incoming voltage (analog) and produces a binary value (digital) for that measurement. The binary value (or code) produced is based on the range and resolution of the ADC. The range specifies the highest and lowest voltages the ADC can measure, and the resolution (as expressed in bits) determines how many binary values or codes can be assigned across that range. The process of converting an analog signal into a digital/discrete one is called quantization.

An example of how an ADC works is shown in Figure 2.9, where the blue line represents a changing voltage signal over time coming into an ADC and the

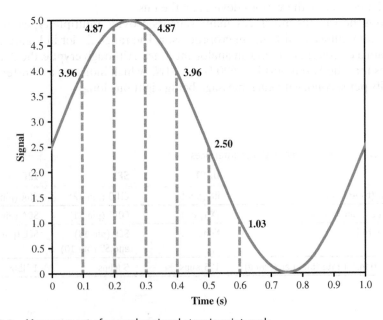

Figure 2.9   Measurement of an analog signal at various intervals.

green lines indicate the various points in time where the signal is sampled. Because the wave is constantly oscillating, the voltage that the device will measure at each time will vary. At 0.1 second, the voltage is 3.96 V. At 0.2 second, it is 4.87 V, and so on. For each of these voltage measurements, the ADC will assign a particular "code" to represent that voltage. If the device has a range from 0 V to 5 V and 10 bits in which to assign codes (providing $2^{10} = 1{,}024$ values), then the voltage 0 V will be assigned the code 0, and the voltage 5 V will be assigned 1,024. All voltages in between will be evenly distributed across this space.

2.11 For the signal in Figure 2.9, what are the maximum and minimum values? At what times do these values occur?

For an example of this process, consider the voltage measurement that occurs at 0.1 second. The result of this measurement is 3.96 V. The ADC will then scale that voltage across the input range to "look up" the appropriate measurement code. This process can be expressed by the equation below:

$$Code = \frac{Sample\ Voltage}{Voltage\ Range} \times 2^{Resolution}$$

For our example ADC with a 5-V range and 10-bit resolution, the resulting code will be

$$811 = \frac{3.96}{5.0} * 2^{10}$$

The result is 811.008 if put into a calculator, but the ADC will "floor" the value to create an integer value of 811. Table 2.3 provides resulting values for several other measurement points on the signal showing the resulting code as an integer and as an unsigned binary value. While we as humans will understand the different codes as integers, the binary value will be the format in which that value

Table 2.3 ADC codes for various sample points

Sample Time (Seconds)	Measurement (Volts)	Integer Code	Binary Code
0.1	3.96	811	11 0111 0001
0.2	4.87	997	11 1110 0101
0.3	4.87	997	11 1110 0101
0.4	3.96	811	11 0111 0001
0.5	2.50	512	10 0000 0000
0.6	1.03	210	00 1101 0010

is physically stored in the microcontroller. For those more interested in learning how to convert information to/from binary, a helpful reference can be found (Mano 2015).

2.12 For an ADC with a range of 10 V and eight-bit resolution, what integer code will be reported for a voltage measurement of 7.34 V?

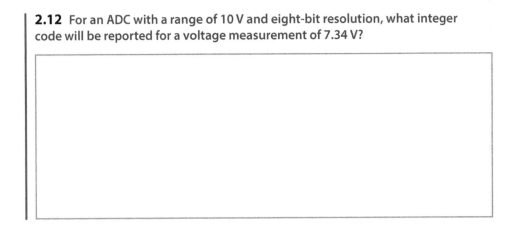

As with any interface, there are limits and design decisions that should be made when utilizing an ADC. First, the input range for the ADC must match the range of voltages coming from the selected sensor. If an ADC has an input range of 0 V to 5 V, it may be damaged if the incoming voltage is outside of that range. Second, the range of the incoming signal should attempt to cover the range of the ADC. If a signal is within 0 V and 5 V but varies only between 2.5 V and 3.0 V, then only a small set of codes can be assigned to that signal. In general, the more codes that are assigned across the signal range, the more information can be captured about the signal. Finally, the range and resolution of the ADC specify the smallest possible signal change an ADC can notice. For a system with a 5-V range and 10-bit resolution, that limit is

$$\text{Limit} = \frac{\text{Range}}{\text{Resolution}} = \frac{5 \text{ V}}{2^{10}} = 0.004 \text{ V} = 4 \text{ mV}$$

This calculation determines how much a signal must change before it is noticed by the ADC. Thus, if the changes experienced by a sensor are less than the limit, the output of that sensor should be amplified to increase the signal throughout the range.

2.13 Would an ADC with a 10-V range and eight-bit resolution be capable of distinguishing between values that are 5 mV apart? Calculate the resolution limit and justify your response.

2.5 Automating Analysis with Programming Languages

The preceding sections discussed how to select sensors (Section 2.3) and then interface them with a microcontroller (Section 2.4). This section focuses on "what to do" with that information once it is on the computer. As discussed in Section 2.3, engineers collect data to "answer questions." In this section, several techniques for answering those questions are explored.

Consider the situation where an engineer has just finished developing the most accurate lens for a new space telescope and is now ready for launch! This sensitive equipment has been built in one part of the country but must be shipped many miles to its destination. The shipping company promises they will take the utmost care of the package. Still, just to be sure and to protect the engineer's hard work, a small accelerometer is included in the box to determine if the lens has been damaged in transit. When the package arrives, a portion of the data analyzed is shown in Figure 2.10.

The data report the total acceleration experienced by the package in terms of *g*'s. One *g* is the force of gravity when standing still. The acceleration will change if something pushes, moves, or drops an object. Examining the results, it appears that

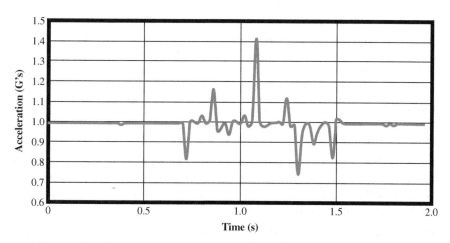

Figure 2.10   Acceleration experienced over time by the package.

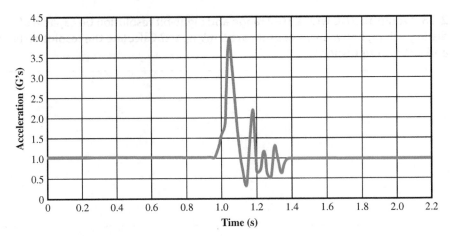

Figure 2.11   Significant acceleration experienced.

the package was subject to some forces (a maximum of 1.4 *g*'s) but nothing that should be hazardous; however, as the engineer continues to examine the data, a later measurement is found, as shown in Figure 2.11.

These data show that the highly sensitive lens was subjected to nearly four times the force of gravity, leading the engineer to consider what happened during shipping and if the device is broken.

The preceding example shows how engineers can employ sensors in typical settings. But in this discussion, what was not covered was how that analysis takes place. One could imagine that an engineer may manually look through some data. Still, if that information is from hours of recordings, it is highly possible that some key information would be missed or that the engineer simply could not examine all of it. Herein is the need for analysis tools such as Excel, MATLAB, and other programming languages (C++, Python, or Java) to automate the analysis of these situations. Consider what could be accomplished if the package would automate this analysis during the trip rather than someone having to examine the data after the fact.

Returning to the example about monitoring a package, it is often important to know the minimum and maximum values in some data stream. To find those values in a stream of data, a simple code pattern can be used to "walk through" the data and find the minimum and maximum values. Figure 2.12 shows a small C++ snippet that will examine every result in an array called sensor readings to find the minimum and maximum values.

A brief description for key elements in the code snippet:

- **Lines 2–3:** Create an array of length 200 to hold the sensor readings from the accelerometer (we assume that some other method places the readings in the array).

- **Lines 6–10:** Create two variables, *min_accel* and *max_accel*, to hold the future minimum and maximum values. They are "initialized" in Line 10 to be the first element of the array. As the code progresses, these variables will be updated. Within a computer, the first element in an array is at position 0.

- **Lines 13–29:** Implement FOR Loop to "walk through" the array. The code will execute these lines in order until the end of the array is reached.

```
1     //an array to hold sample readings
2     const int num_samples=200;
3     float sensor_readings[num_samples];
4
5     //variables to hold minimum and maximum acceleration
6     float min_accel;
7     float max_accel;
8
9     //initialize the minimum and maximum values
10    min_accel=max_accel=sensor_readings[0];
11
12    //loop through all sensor values
13    for (int i=0;i<num_samples;i++)
14    {
15
16       //pull the most recent sample from the array
17       float acceleration = sensor_readings[i];
18
19       // see if the current value is above our maximum
20       if (acceleration > max_accel)
21       {
22          max_accel = acceleration;
23       }
24
25       // see if the current value is below our minimum
26       if (acceleration < min_accel)
27       {
28          min_accel = acceleration;
29       }
30    }
```

Figure 2.12 Finding the minimum and maximum values in an array.

- **Line 17:** The "*i*th" element of the array is stored as the variable *acceleration*. The *i*th element will start at "0" and then increment to "1" on the next loop iteration. In this way, all array elements are accessed in order.

- **Lines 20–23:** Perform a comparison to see if the latest value is larger than the maximum. If so, store the result.

- **Lines 26–29:** Perform a comparison to see if the latest value is larger than the minimum. If so, store the result.

The result of this code is that once it is run, the variables *min_accel* and *max_accel* will contain the minimum and maximum acceleration experienced by the sensor. Overall, this automated approach will be much faster and reliable to implement than an engineer examining the sensor data by hand.

While knowing the minimum and maximum value from some data stream is important, other statistics, such as *mean* (μ) and *standard deviation* (σ), can be calculated that would provide more insight into what has occurred. The mean (also called an average) and standard deviation can be calculated by the equations shown below, where N is the number of samples and x_i is an individual sample in the set:

$$\mu = \frac{1}{N}\sum_{i=1}^{N} x_i \qquad\qquad \sigma = \sqrt{\frac{\sum_{i=1}^{N}(x_i - \mu)}{N}}$$

Specifically, the mean and standard deviation describe the "overall" experience from the data beyond simply the minimum and maximum values that provide a singular value. For example, in analyzing the data in Figure 2.11, the average acceleration is 1.07 g, which does not indicate a significant acceleration event. However, the standard deviation of 0.42 g shows that while the "average" appears typical, the sensor experienced a wide range of high g-forces during the shipment.

Implementing an average within code can be easier than the thresholding analysis shown previously.

Much of the code in Figure 2.13 is like the previous example; however, a new variable to hold the sum of the array (Line 6) is created, and rather than comparing the latest acceleration to a minimum and maximum value, it is simply added to the running total (Line 12). In one distinction, the result for the average must be

```
1    //an array to hold sample readings
2    const int num_samples = 200;
3    float sensor_readings[num_samples];
4
5    //hold the sum of the array
6    float sum = 0;
7
8    //loop through all sensor values
9    for (int i = 0; i < num_samples; i++)
10   {
11       //pull the most recent sample from the array
12       float acceleration = sensor_readings[i];
13
14       //add the most recent value to the sum
15       sum = sum + acceleration;
16   }
17
18   //divide the sum by the length
19   float average = sum / (float)num_samples;
```

Figure 2.13   Implementing an average calculation.

calculated after the FOR loop (Line 19). An aspect of the C++ language requires that the key word *float* be used in the division to ensure that the result is not truncated to an integer value and that precision is not lost.

> **2.14** For the following set of values, calculate the average and standard deviation: 2, 6, 29, −5, 1, 7, and 9.

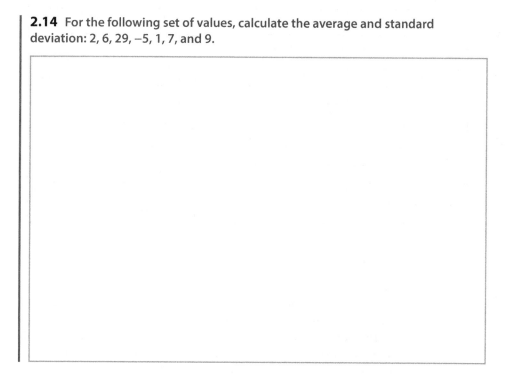

It is important to note that while tremendously complex, computers are simply devices that follow instructions provided by humans. The particular "language" that humans use to instruct these machines are programming languages such as C/C++, Java, Python, and myriad others. However, fundamental to all those languages, each program is translated into a series of small instructions that the computer executes. These instructions are typically low-level operations, such as "add these numbers," "move this data over here," "compare this value," and so on. Much of this translation process is hidden from the user, but it can have significant impacts on performance. For example, the "simple" FOR loop in Figure 2.13 may result in hundreds of instructions to the computer, whereas the human only had to write several lines. The complete discussion of how computers operate and are designed is beyond the scope of this work, but an interested reader is directed to Patterson (2020).

2.6 Communicating Data and Results via RF Links

For the final consideration of the Data Acquisition pipeline, there are often situations where the information needs to be sent "away" from the system conducting the sensing. As seen in Figure 2.1, the data from the sensors can be either stored locally on the microprocessor or sent externally to the Cloud for storage. This latter option is helpful, as it allows anyone to see the latest data from the sensor without physically accessing it. Consider if a DAQ system was built to monitor remote environmental conditions. It would not be beneficial (but it might be fun!) to travel there when new measurements are available.

Specifications

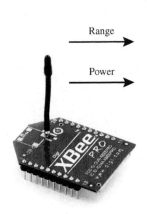

Range →

Power →

Specification	XBee	XBee-PRO
Performance		
Indoor/Urban Range	Up to 100 ft (30 m)	Up to 300 ft (90 m), up to 200 ft (60 m) International variant
Outdoor RF line-of-sight Range	Up to 300 ft (90 m)	Up to 1 mile (1600 m), up to 2500 ft (750 m) International variant
Transmit Power Output (software selectable)	1 mW (0 dBm)	63 mW (18 dBm) 10 mW (10 dBM) for International variant
RF Data Rate	250,000 bps	250,000 bps
Serial Interface Data Rate (software selectable)	1200 bps – 250 kbps (non-standard baud rates also supported)	1200 bps – 250 kbps (non-standard baud rates also supported)
Receiver Sensitivity	–92 dBm (1% packet error rate)	–100 dBm (1% packet error rate)
Power Requirements		
Supply Voltage	2.8–3.4 V	2.8–3.4 V
Transmit Current (typical)	45 mA (@ 3.3 V)	250 mA (@3.3 V) (150 mA for international variant) RPSMA module only: 340 mA (@3.3 V) (180 mA for International variant)

Figure 2.14   XBee Series 1 radio and data sheet.
Source: © SparkFun Electronics.

Any data sent "away" from the microprocessor will be done with a radio. These devices operate by encoding information within light waves that are far outside the visible spectrum. Thus, while we are all awash in radio waves every day, we cannot always see this information around us. Many radio technologies are familiar in daily life: simple remotes to control a TV, Bluetooth between smartphones and audio headsets, and Wi-Fi to connect devices to the internet. All these technologies are based on similar principles. When selecting parts for a project, one must consider the effective range for the devices, their transmission power, and how interference with the environment will play a role.

To illustrate these points, a small radio and its data sheet are shown in Figure 2.14. The XBee radio is a cheap and easy-to-use component that works with many microprocessors. As discussed in Section 2.3, a data sheet provides many key parameters that describe a device's performance. Notably, this data sheet describes the parameters for two distinct models: a baseline XBee and an upgraded XBee Pro model.

2.15 Consider a radio-based device that you utilize in daily life. What is its purpose? What is its effective range?

Examining the range for the radios, several variations are possible based on the environment that is used. For the XBee model, when used "outdoors," the range is up to 300 feet but that range drops to 100 feet when used "indoors." The descriptors for indoor and outdoor use are not well defined, but the primary outcome is that when obstacles are in the way of the radio, they cause interference, and the radio's effective range is less. This may be a familiar shared experience when attempting a cell phone call indoors versus outdoors.

The Pro model reports slightly higher values for the indoor and outdoor ranges with 300 feet and one mile, respectively. This model achieves a greater range by overcoming those obstacles with increased transmission power. The Pro version uses 63 mW for transmission, while the base model uses only 1 mW. While the greater transmission range might be desirable, the extra energy required will more quickly drain any battery source the radio is connected to.

References

Internet Engineering Task Force. (2022). "Terminology, Power, and Inclusive Language in Internet-Drafts and RFCs." datatracker.ietf.org/doc/html/draft-knodel-terminology-10.

Mano, M. M., Kime, C. R., and Martin, T. (2015). *Logic and Computer Design Fundamentals*. 5th ed. London: Pearson.

Miller, Dean J., et al. (2020). "Analyzing Changes in Respiratory Rate to Predict the Risk of COVID-19 Infection." *PLoS One* 15, no. 12 (2020): e0243693.

National Institute of Standards and Technology. (2021). "Guidance for NIST Staff on Using Inclusive Language in Documentary Standards." nvlpubs.nist.gov/nistpubs/ir/2021/NIST.IR.8366.pdf.

Patterson, D., and Hennessy, J. (2020). *Computer Organization and Design*. RISC-V ed. New York: Morgan Kauffman.

Chapter 3

Structures and Society: Structural Engineering Problem-Solving Techniques to Design Long-Lasting Solutions

Objectives

Chapter 1 introduced topics about forces acting on materials. This chapter extends the analysis of forces to design simple structures based upon static equilibrium. Structural engineering problem-solving techniques are applied to solve for the state of static equilibrium for simple structures. The simple structures in this chapter include real-world examples, which reinforce the notion that structural engineering design is not undertaken in a void. Instead, it is informed by human needs and is inherently constrained by sociocultural, economic, and environmental factors.

Student Learning Objectives

- Explain how civil and structural engineering shape human culture.
- Explain the structural engineering design process.
- Construct force systems and engineering systems.
- Solve simple engineering mechanics–statics problems.
- Design structural engineering solutions that consider sociocultural, environmental, and economic contexts.

This chapter introduces structural engineering topics that balance structural engineering problem-solving techniques with broader societal considerations. Structural engineering is a subdiscipline within civil engineering. Civil engineers can find themselves practicing in many divergent subdisciplines, yet they all share a common trait in leveraging natural resources to create structures or structural systems with specific functions that meet human needs. These human needs are diverse in nature: from domiciles and places of refuge to places of worship and well-being to places of commerce and recreation and the wide-ranging infrastructure networks connecting all those places.

This chapter offers an introduction to structural engineering concepts and equations that are used to calculate internal forces and deflections in simple structures. You will explore how the selection of different materials impacts simple structural designs, which will begin to develop your insight on the structural engineering design process. The topics introduced in this chapter will extend into future courses included in a typical structural engineering curriculum, including engineering mechanics (statics), mechanics of materials, structural analysis, and structural engineering design (e.g., masonry, reinforced concrete, steel, wood).

3.1 Introduction

Civil engineering is the oldest engineering profession, with a rich history of structures innate to many early human cultures in Africa, the Americas, Asia, Europe, and Oceania. Yet how might we define human culture? Many students studying structural engineering might be compelled to point to the still-standing Great Pyramids in Egypt (Figure 3.1), Mesoamerican cities in Mexico (Figure 3.2), the Great Wall of China (Figure 3.3), or megalithic temples and burial sites in Malta (Figure 3.4) as instances in which human societies first established across different regions of the world. In other words, very old structures are evidence of early human civilization. Yet an anthropologist might encourage you to broaden your thinking!

Figure 3.1 The Great Pyramids of Giza in Egypt were built during the Fourth Dynasty of the Old Kingdom of ancient Egypt between 2600 BC and 2500 BC.
Source: iStock.com/WitR.

Figure 3.2 The ruins of the Palenque city-state. Palenque was built in Mexico in 226 BC to AD 799.
Source: Zoonar GmbH/Alamy Stock Photo.

Figure 3.3 The Great Wall of China was built in discrete segments as early as 220 BC to 206 BC and later joined together.
Source: iStock.com/Zhaojiankang.

Figure 3.4 The Megalithic Temples of Malta were built during three distinct time periods between 3600 BC and 2500 BC.
Source: iStock.com/Aksenovko.

Anthropologists are people who study human behavior, cultures, and societies in both the present and the past. One prominent anthropologist, Margaret Mead (Figure 3.5), is attributed to an argument that the earliest sign of true human civilization is a healed femur, or leg bone. When animals are injured, they become immobile, making them susceptible to predators and death. Typically, animals are unable to care for injured members in their herds or packs, as it requires expending precious time and energy that they must conserve for themselves in their own plight for survival in the wilderness. Human civilization, according to Mead's argument, started when our earliest human ancestors tended to each other's significant

Figure 3.5   Anthropologist Margaret Mead (1901–1978) is often attributed to an argument that healed femurs, or leg bones, are notable indicators of early human civilization.
Source: Smithsonian Institution/Wikimedia Commons.

injuries, like broken bones. A broken bone, like a femur, requires extensive time to heal, and discovering healed femur bones at archaeological sites is indicative of ancient human clans expending precious time and energy to tend to one of their injured clan members. That altruistic behavior is a distinctly human trait and can point to the establishment of our earliest human civilizations. The oldest, healed femur bone has been dated at 15,000 years old!

Many anthropologists will also point to unearthed structures like simple tools and simple weapons as instances of early human civilization first taking root in a region. To that end, archaeologists have discovered 70,000-year-old spearheads in the Tsodilo Hills of Botswana!

For our purposes in learning about structural engineering, it is worth observing that our early human ancestors had unmet needs that required structural engineering solutions. The human skeletal system, for example, is a structural system itself (Figure 3.6). It sustains external loading when we walk upright, and it distributes that external loading into our bones, joints, and muscles as internal forces. Our early ancestors must have devised splints utilizing natural materials to set and heal broken bones. Sharpened arrowheads, as another example, come from natural materials, and their shape is modified to achieve a specified function for hunting and defense. In a way, our earliest human ancestors had insight into structural engineering phenomena in order to create these types of solutions; in essence, our human ancestors were innately structural engineers! If you are reading this, then you are likely human yourself, meaning that you, too, hold enough innate intuition to master structural engineering design just like our ancestors.

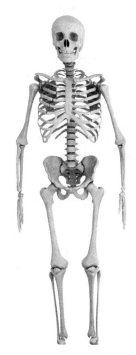

Figure 3.6   The human skeletal system is a structural system.
Source: iStock.com/Abidal.

3.1 Reflect on your own childhood and family upbringing. What is the earliest instance in which you can remember engaging in structural engineering–like activity (i.e., assembling toys, building a doghouse, digging postholes, setting a broken bone)?

3.2   A Brief History of Structural Engineering

Structural engineering has come a long way since the Neolithic and pre-Columbian eras. In antiquity, people holding privileged positions in society would be trained to use mathematics (e.g., geometry and algebra) and physics to design structures *to the best of their knowledge.* This knowledge was gained from their own elders in master–apprentice working relationships. When a structure failed or collapsed, any insight that could be gained from that failure would be passed on to their apprentices. The apprentices would someday become masters themselves and design new structures that incorporated this updated knowledge. This approach to structural engineering design was highly *empirical*, meaning that the knowledge was gained from simple observation or simple experimentation. This approach is highly limiting in designing structures that are safe and resilient against natural disasters like earthquakes, high winds, or flooding. After all, a master would be able to learn about a structure's inability to withstand those types of loads only *after* it fails!

3.2 Reflect on knowledge that your *family* holds and passes down from generation to generation. Is there an activity, practice, or skill that has been passed down to you (e.g., artistry, baking or cooking, painting, quilting, woodworking)? How is that activity *empirical* in nature? Have there been instances where you learned new knowledge (perhaps through failure) and intend to pass it on to other persons in the future?

While there were undoubtedly varying structural engineering design approaches throughout Africa, Asia, and the Americas by the 1400s, it was the Italian Renaissance in the 15th and 16th centuries that ended up greatly shaping the structural engineering process as we understand it in Western society today. Galileo Galilei (Figure 3.7) established principles about the *inherent strength of materials*. Later, Robert Hooke (Figure 3.8) established relationships between *force and displacement*. Charles Coulomb (Figure 3.9) extended that force–displacement relationship toward an understanding of *linear elasticity*. It is on their collective, foundational contributions that classical structural theories emerged (e.g., Lamé, St. Venant, Castigliano, Mohr, Müller-Breslau) throughout the 19th and 20th centuries that approached structural engineering design not as an empirical practice but as a *scientific* process with approximate methods. In the mid- to late-20th century, the slope-deflection method and matrix analysis helped structural engineers solve complex, statically indeterminate structures by hand computation (Figure 3.10). The advent of computers in the late 20th century through the present has opened the door to solving even more complex structural engineering problems using finite element analysis (FEA) (Figure 3.11).

Figure 3.7 Galileo Galilei (1564–1642).
Source: Justus Sustermans/Wikimedia Commons.

Figure 3.8 Robert Hooke (1635–1703).
Source: Mary Beale/Wikimedia Commons.

Figure 3.9   Charles Coulomb (1736–1806).
Source: Hippolyte Lecomte/Wikimedia Commons.

Figure 3.10   The Empire State Building in New York City was built from 1930 to 1931, using hand calculation to inform its framed structure design.
Source: David Ball/Alamy Stock Photo.

Figure 3.11 The Burj Khalifa in Dubai, United Arab Emirates, was built from 2004 to 2009 and is the tallest building in the world. Its structural design was informed through advanced construction materials, advanced construction methods, and finite element analysis (FEA).
Source: MB_Photo/Alamy Stock Photo.

The nature of learning about and practicing structural engineering in the United States shifted in the mid-1800s from a master–apprentice model to a licensure model, with national engineering societies like the American Society of Civil Engineers (ASCE) being founded to inform licensing bodies in each state about the appropriate credentials a person must hold in order to be recognized and legally allowed to practice as a structural engineer in that state. In many states, the minimum requirements include holding an ABET-accredited degree in engineering or civil engineering as well as having successfully passed specified exams (e.g., the Fundamentals of Engineering [FE] exam, Principles and Practice of Engineering exam) and practiced as a professional engineer under a licensed structural engineer for a specified number of years. As a first-year student, you should consult with your civil engineering faculty and/or adviser about the FE exam so that you can work toward securing an Engineer-in-Training (EIT) license in your state.

3.3 The Structural Engineering Design Process

The structural engineering design process is similar to the broader engineering design process. A structural engineer gathers information from relevant sources to solve a specific problem for a particular situation. For structural engineers, the structural engineering design process might be simplified to three overarching steps:

1. Assessing the structural demand
2. Establishing the structural capacity
3. Safeguarding the structural performance

Step 1: Assessing the Structural Demand   Structural systems are made up of structural members that interact with their environment on Earth (and, in an increasing number of instances, in space). Structural systems on Earth interact with their local environment in a number of ways. Structural engineers must understand a structural system's dead and live loads, and they must understand how those external forces flow through the structural system using a load path analysis.

Dead Loads (DL)   Dead loads, or *gravity loads*, are associated with *permanent* or *semipermanent loading* acting on a structure for its entire service life. Dead loads can include the structure's self-weight and the weight of mostly immovable, heavy objects inside (e.g., heavy equipment in a medical facility). A wooden structure will have less dead load acting on it than a steel structure, attributable mostly to the disparate material density of each material (i.e., wood is less dense than steel). Yet this example is true only if the volume of material between the two are equal. Thus, the volumetric shape of each member also influences the extent of the dead load. For example, a very large wooden column will generate larger dead load values in a structural system than a very slim steel cable. Structural engineers will often *idealize* a complex structure into lines and opt to conduct calculations using *linear feet* instead of *volumetric dimensions*. The dead load for a *linear* member in a structural system (e.g., beam or column) can be calculated by taking the product of the unit weight, γ, of the structural material and its *cross-sectional area, A*. The resulting value is a dead load distributed over a linear distance, as shown in the following equation:

$$DL = \gamma A \qquad [\text{force/distance}]$$

It is important to note that the unit weight of materials (force/volume) is subtly distinct from the density of materials (mass/volume) in that unit weight accounts for the rate of acceleration of gravity.

Live Loads (LL)   Live loads, or *service loads*, are associated with *temporal loading* that acts on the structure for relatively brief instances of time and can vary in their magnitude, orientation, and location. Human foot traffic on structures is considered a live load, and live loads can also include vehicular traffic and reconfigurable office furniture or equipment. Weather phenomena (e.g., wind, earthquakes, snow) also act temporally and are considered live loads. Typically, live loads act on *areas* (e.g., on a floor or on a wall), resulting in values with units of force/area. Specific values of live loads or service loads can be found in a variety of tables, most notably in the ASCE Structural Engineering Institute (SEI) 7 *Minimum Design Loads and Associated Criteria for Buildings and Other Structures*.

Load Path Analysis   The summation of the DL and LL constitutes the total load acting on the structural system. Yet the LL acts on floors or walls, meaning that our current definition of the total loading has external loading acting only on floors or walls. We have not yet fully accounted for how that total loading will transmit from the floors to the foundation of the structure. A load path analysis is conducted to understand fully how total loading transmits from the floors to underlying floor beams, beam girders, columns, and, finally, the foundation. The *flow of forces* from the external loading to the foundation transmits through the structure as *internal forces*. This advanced load path analysis is conducted for complex structures in structural analysis courses and is not detailed in this chapter.

3.3 Lumber is a construction material used in many structural systems, like residential homes and low-rise commercial structures. Lumber is sourced from a variety of tree species, which the American Wood Council (awc.org) has cataloged in their *National Design Specification for Wood Construction with Commentary*. Consider an eight-foot-long piece of "2 × 4" pine (pine, Southern yellow). Sketch the actual dimensions of the "2 × 4" (Hint: It is not 2.0 inches by 4.0 inches) and calculate the gravity load associated with this member in a larger structural system. Report your answer in lbf/foot. Assume the unit weight is approximately 50 pcf.

3.4 Multistory school buildings contain a variety of spaces in which human foot traffic and configurable furniture (e.g., desks) can impart live loads on each floor of the structure. ASCE/SEI 7 states that the live load for classrooms should be 40 psf, the live load for first-floor corridors should be 100 psf, and the live load for corridors above the first floor should be 80 psf. Determine the dimensions of the classroom that you are situated in and sketch a detailed floor plan of the nearby classrooms and corridors to the nearest inch. Calculate the service load (in units of lbf) for the floor plan you have sketched. Additionally, speculate why corridors above the first floor have a reduced live load value.

Step 2: Establishing the Structural Capacity   The selection of a live load (LL) depends on the intended function of the structure itself. The function of the structure may begin to influence what types of materials are used in the structure. A variety of complementary constraints inform this material selection process, including where the structural materials originate, the associated transportation costs to deliver materials to the construction site, the calculated mechanical response of the structural materials due to the applied loading, and material durability issues.

Sourcing Materials   First, the selection of construction materials is driven by the availability of the materials themselves. Construction lumber is abundant in forested areas but is limited in arid and semiarid regions, like deserts. Masonry (e.g., clay bricks, cinder blocks) and ordinary Portland cement concrete can be manufactured locally given a sufficient supply of raw ingredients (e.g., lime-rich stone, silica-rich clay) and potable water. Other structural materials, like steel and aluminum, can be manufactured locally if the raw ingredients (i.e., iron ore and bauxite, respectively) are available. The extraction of any of the raw, virgin materials needed to manufacture the construction materials holds consequential implications for the natural environment and must be carefully considered when sourcing raw materials. When possible, reusing reclaimed construction materials in new construction can greatly diminish the environmental impact (e.g., using recycled aggregates in new concrete construction).

Transportation Costs   Often, the most appropriate construction materials for a specific structural system may not be locally available and must be imported from outside regions or foreign countries. A significant amount of construction-grade lumber is imported from Canada to the United States, and a significant amount of construction-grade steel is imported from China to the United States. Material and transportation costs can fluctuate significantly in the globalized economy, which can lead to cost-prohibitive material selections. For example, transportation engineers who specialize in pavement engineering closely monitor the costs of asphaltic cement concrete (ACC) and Portland cement concrete (PCC) to decide which structural material will be used to construct roadways. Moreover, it is important to recognize that material selection in any engineering design process is closely intertwined with broader economic issues in our globalized economy, environmental issues in burning fossil-based fuels to traverse large distances and manufacturing the structural members themselves, and geopolitical tensions in instances where critical commodities (e.g., oil and raw materials) are sourced from authoritarian regimes that conflict with the United States or democratic interests.

Mechanics of Materials   Once a suite of appropriate materials is identified, an inventory of the *material engineering properties* is used to calculate the *mechanical response* of each structural member subjected to the predicted loading from a load path analysis. A future course (mechanics of materials) employs advanced analysis to calculate how the *material engineering properties* (like Young's modulus, E, fracture strengths, yield strengths, and so on) can be used to influence the extent of deformation attributable to internal uniaxial forces, internal shear, internal bending, and internal torsion. One simple example of a mechanics of materials

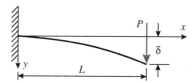

Figure 3.12 An illustration of an idealized cantilever beam.

problem is that of a cantilever beam (Figure 3.12), whereby the end deflection, δ, is calculated using

$$\delta = \frac{PL^3}{3EI}$$

where

    P is a concentrated point load applied at the free end of the beam (force),
    L is the span length of the cantilever beam (distance),
    E is Young's modulus for the material (force/area), and
    I is the area moment of inertia for the cross-sectional shape of the member (distance^4).

The area moment of inertia, I, is an engineering measurement for the *tendency of a structural member to resist bending* and is calculated using the *cross section* of the structural member subjected to bending. For a structural member with a circular cross-sectional shape subjected to bending, the area moment of inertia for *that specific, circular-shaped structural member* can be calculated using

$$I_{\text{circle}} = \frac{1}{4}\pi r^4$$

where r is the radius of the circle (distance).

There are many other cross-sectional shapes that structural engineers use for structural members. Another *very common shape* is a rectangular cross section. Yet a structural member with a rectangular cross section will behave *very differently* depending on how bending occurs. Consider a simple ruler (Figure 3.13). When positioned flat, the ruler can bend easily by hand manipulation, as seen in Figure 3.13. When positioned upright and bent along its other dimensional axis, the ruler will be *more resistant to bending*. The material itself has not changed, yet the cross-sectional shape has in a way! The area moment of inertia, I, for a rectangular cross section can be calculated by carefully considering which dimensional parameter is acting as the width of the cross section, b, and which is acting as the height of the cross section, h. It is important to realize that *depending on which direction bending is occurring*, the parameters b and h can shift! The area moment of inertia, I, for a rectangular cross section can be calculated using

$$I_{\text{rectangle}} = \frac{bh^3}{12}$$

where

    b is the width of the rectangle (distance) and
    h is the height of the rectangle (distance).

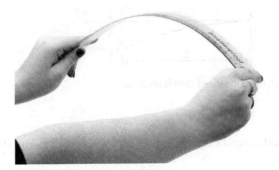

Figure 3.13   A ruler will more easily bend in one direction than another.
Source: Art Directors & TRIP/Alamy Stock Photo.

Figure 3.14 illustrates a typical reference for how a structural engineer might find the area moment of inertia for a rectangular cross section in a textbook on the mechanics of materials. The structural engineer must be careful to understand *which* of the two equations applies to the calculation at hand.

More advanced analysis regarding force and displacement is introduced in courses on the mechanics of materials (or strength of materials); it is not covered further in this chapter.

Durability   Finally, structural engineers work closely with materials engineers to select construction materials that are sufficiently durable and can withstand the environment that they are in for decades to come. Material durability is defined as the material's ability to maintain its integrity for many decades, resisting the tendency to decay, corrode, or dissolve. For example, lumber is a natural material subject to material damage from rot, fungal disease, and wood-boring insects. If this material damage becomes appreciable, then the structural member loses its mechanical ability to withstand its designed loading, putting the structural system at risk of failure. In this example, structural lumber performs best when its internal moisture content is very low and it has been treated with insect-repelling chemicals. Steel, as another example, will actively corrode when exposed to an electrolyte (e.g., water or salt water) in a process called *rusting*. Rusting is a chemical process that erodes the strong base material, leaving behind weak, oxidized material that readily

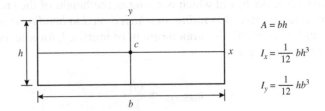

$$A = bh$$

$$I_x = \frac{1}{12}\,bh^3$$

$$I_y = \frac{1}{12}\,hb^3$$

Figure 3.14   The area moment of inertia for a structural member with a rectangular cross section can be calculated using *b* and *h*. Note that the parameters can flip depending on whether the structural member is being subjected to bending about one axis (e.g., *x*-axis) or the other (e.g., *y*-axis).

flakes away. The irreversible loss of this structural material significantly reduces the cross-sectional area of the structural member. Thus, the structural capacity of the structural member is diminished, leading to the risk of structural failure. A structural engineer might need to design for additional measures to ensure the durability of the wooden structure or steel structure if the structural system is situated in a moist climate, a marine environment, or a region with particularly invasive insects or fungal disease.

3.5 Consider your current classroom or building. Inventory the construction materials that you can observe (e.g., clay brick, cinder block masonry, reinforced concrete, lumber, steel, glass). Based on any relevant contextual information you can find about the structure that you are in (e.g., when it was built), speculate on where each of your inventoried construction materials likely originated from. Did any of the materials come from the municipality itself? From the state? From elsewhere in the country? From elsewhere in the world?

3.6 Locate two rulers so that you can use one ruler to measure the other. Measure the dimensions of the cross-sectional area of one of the rulers to the nearest millimeter. Calculate the area moment of inertia, I, for its two orientations (e.g., x and y) when subjected to bending in the x- and y-directions. Which value is higher? Which value is lower? How do these values correspond to your ability to physically bend the ruler along the x- and y-axes? Comment on your effort.

3.7 Students at many colleges and universities enjoy socializing at homes and apartments that have *cantilever*-style balconies. A cantilever is a simple structural member that is firmly affixed at one end and has an external load imparted along its length and/or at its end. It can become extremely dangerous when these balconies are overloaded with too many people, whereby the end deflection, δ, becomes very large and the internal forces become larger than the ultimate strength of the material, resulting in catastrophic failure. Consider a cantilever beam, *AB*, made up of an eight-foot pine (Ponderosa) "2 × 4" with an elastic modulus, *E*, of 7.5 ksi. Calculate the end deflection, δ, if the end of the beam is subjected to a concentrated point load, *P*, of 60 lbf. Note that the units of ksi and psi are related in that psi is pound-force per square inch and ksi is kips per square inch. One kip equals 1,000 lbf.

3.8 Consider the previous problem of the cantilever-style beam *AB*. If the material changed from pine (Ponderosa) to steel (E_{steel} = 29 × 10³ ksi), calculate the revised end deflection of the cantilever beam. Is the calculated value better or worse for a structural engineer? What other factors must the structural engineer consider in deciding between lumber or steel for this specific structural scenario?

Step 3: Safeguarding the Structural Performance The structural engineering design process is highly iterative. When conducting an initial analysis, a structural engineer *estimates* the DL values based on the initial material selection and initial material shape. This initial analysis allows the structural engineer to calculate critical values (e.g., end deflections), which allows them to compare the values to acceptable deflection limits and the strength of the material. If those limits are exceeded, then the structural engineer has an unlimited number of options to consider in their iterations of the structural design: (1) the assemblage (e.g., triangular truss, square framed portals) of structural members themselves in the structural system, (2) the selection of any number of construction materials for any of the structural members, and (3) the dimensions of any of the individual structural members themselves. In any of these iterations, a structural engineer is always assessing three critical outcomes that they are responsible for.

Primary Responsibility The primary responsibility of any engineer is to practice only within the field that they are specialized in. A structural engineer is specialized to optimize the design and construction of structures such that they are *safe* for its users and stakeholders. When subjected to anticipated loading (e.g., earthquakes, explosions, flooding, hurricanes, tornadoes), the structure must remain standing long enough for its users to evacuate safely. To reinforce this standard of safety, structural engineers introduce significant redundancy into structural systems such that the *factor of safety* (*FOS*) or *safety factor* (*SF*) is approximately 2 to 3 at every structural member and joint. These factors are calculated using the simple equation

$$FOS \text{ or } SF = \frac{\text{Capacity}}{\text{Demand}}$$

where
  Capacity is the strength of the structural member and
  Demand is the total force in the worst-case scenario (e.g., earthquake and wind loading) predicted to flow through that member (e.g., LL + DL).

Using the *FOS* or *SF*, in the case when one (or several) member(s) or joint(s) fail(s), the rest of the structure has sufficient capacity to absorb the sudden redistribution of forces in the structural system, safeguarding against catastrophic failure or collapse. The *FOS* or *SF* can also be applied for deflections, like the end deflection of a cantilever beam.

Secondary Responsibility The structural engineer's primary responsibility toward safety has been criticized as being insufficient for our modern-day civilization. After all, what is the point of safely evacuating a structure during a disaster if the structure is rendered inhabitable or deemed unsafe afterward, necessitating its demolition? In 2011, a powerful 6.3-magnitude (Richter scale) earthquake struck the Canterbury region of New Zealand near Christchurch. The earthquake resulted in 158 deaths, and caused an estimated $40 billion worth of structural damage throughout the region. Many structures in the region were designed using the most up-to-date and appropriate structural engineering design codes of the time, significantly reducing the loss of life (i.e., the structures were "safe enough"). Yet a vast number of those buildings needed to be demolished, as they were not salvageable after the earthquake. This event was a milestone in global structural engineering design standards, where an emergent secondary responsibility for structural engineers is both to design structures to be safe for its users and to design structures that can be *serviceable* and *operational* after extreme loading events. This secondary responsibility has opened up new structural engineering design methodologies in creating *resilient* structures. Resiliency in structural engineering is defined as the ability of a structure to recover its function quickly after an extreme, damaging loading event. One example of how structures are designed to be resilient is that structures built in earthquake zones will contain *sacrificial elements* that intentionally absorb significant energy during extreme loading, which greatly dissipates structural damage and can be easily replaced after the event has concluded.

Responsibility for Sustainability   Structural engineers are humans living in a globalized society, with a responsibility to steward the Earth's precious natural resources to meet civilization's present needs. Structural engineers can opt for many decisions throughout the structural engineering design process to promote the positive *sustainability* of the human world we live in and the natural environment. Sustainability is defined as an ability to meet the needs of the present without compromising the ability of future generations to meet their own needs. Structural engineers can opt to design structures that are less energy intensive (e.g., use natural light for interior lighting needs) with structural materials that are renewable and locally sourced (e.g., wood) and designed with long service lives due to their durability. The United Nations adopted 17 Sustainable Development Goals that speak toward broader sustainability issues that structural engineers are increasingly considering in their engineering design and practice.

3.9 In your own words, explain the three responsibilities of a structural engineer.

3.10 Consider the problem of the *cantilever*-style beam *AB* that uses steel. Assume that your calculated end value is the *maximum allowable value* before failure occurs. If a safety factor of 2.5 is applied, then what is the recommended limit on the concentrated end load for the structural design?

3.11 Conduct research on at least one energy-absorbing technology or sacrificial member used in structural systems (e.g., lead-rubber bearings, steel plate shear walls, controlled rocking, tuned mass dampers, seismic cloaking). In your own words, summarize your understanding of your selected system and its application to structural engineering design.

3.12 Conduct research on the United Nations' 17 Sustainable Development Goals. In your own words, explain how at least one of these goals relate to structural engineering design processes.

3.13 Conduct research on the typical shape of *truss structures* and *framed structures*. Differentiate the similarities and differences of their defining characteristics in a Venn diagram.

3.14 Sketch two two-dimensional drawings of a 50-foot-wide and 100-foot-tall office building. In your first sketch, use *truss* structural members in your design where no individual member exceeds 14 feet. In your second sketch, use *truss* structural members in your design where no individual member exceeds seven feet. Compare and contrast your designs. Speculate on specific scenarios where seven feet and 14 feet are appropriate limits for disparate structures.

3.15 Sketch two two-dimensional drawings of a 50-foot-wide and 100-foot-tall office building. In your first sketch, use *framed* structural members in your design where no individual member exceeds 10 feet. In your second sketch, use *framed* structural members in your design where no individual member exceeds 20 feet. Compare and contrast your designs. Speculate on specific scenarios where 10 feet and 20 feet are appropriate limits for disparate structures.

3.4   Problem-Solving Techniques in Structural Engineering

Structural engineers must be able to analyze simple structures using hand calculation to verify the veracity of structural engineering analysis software.

The first step in analyzing a simple structure is *idealizing* any complex, three-dimensional structure into a simplified, two-dimensional representation made up of lines and nodes. Consider a large aircraft (Figure 3.15) where the aircraft wings extend away from the fuselage (i.e., aircraft body). To the untrained eye, this three-dimensional structure is highly complex, with windows and landing gear and rounded surfaces, among many other features. To a structural engineer, however, this complex structure can be *idealized* into a simpler structure in order to conduct a preliminary analysis of the internal forces within the structure.

An aircraft made up of a fuselage and two wings might be idealized from a top-down view, as shown in Figure 3.16. In this idealized drawing, we see that the

Figure 3.15   Large aircraft midflight.
Source: Brad Striebig.

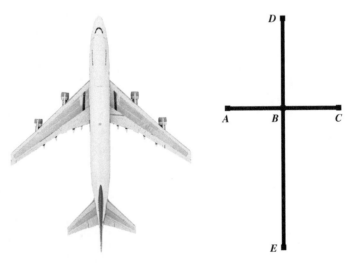

Figure 3.16   A top-down view of an idealized aircraft.
Source: iStock.com/Naypong.

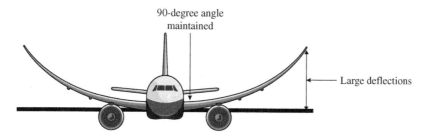

90-degree angle
maintained

Large deflections

Figure 3.17 A depiction of the aircraft wing maintaining its original angle at the juncture with the fuselage regardless of the deformation occurring elsewhere along the aircraft wing.

fuselage is represented by line DBE and that the two wings are represented by line segments AB and BC. Various joints are displayed throughout and can communicate critical information. Interestingly, joint B is communicating information about *how* the line segments DBE, AB, and BC interact. The line segments meet at the square symbol, which indicates that the angles between the lines *must be maintained* in the event of tolerable deformations. That is, the 90-degree angle that the three line segments make with each other must be maintained when we conduct our structural engineering analysis. In other words, the structural engineer acknowledges the fact that the aircraft wings do not hinge or pivot at the juncture with the fuselage. The 90-degree angle is maintained when the wings are deformed, as seen in Figure 3.17.

Not all structures have their structural members meet at connections or joints that maintain the original angles. There are other connection types that intentionally *allow for very small rotation* (Figure 3.18). Unlike the *fixed* connection we previously considered, these connections are called *hinges*, *internal pins*, *pins*, or *rollers*, and they allow for the angle between structural members to change slightly. A structural engineer would idealize these hinge connections using small circles adjacent to black squares or other symbols to communicate that there is an ability for the structural members to rotate slightly (Figure 3.19).

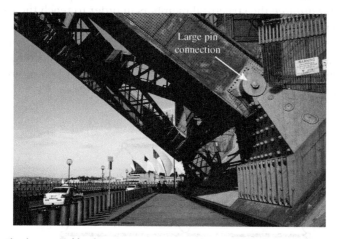

Large pin
connection

Figure 3.18 This large steel bridge structure is an example of how structural members can be joined through a very strong *pin* connection. The connection type allows for small, angular rotation.
Source: © Martin Pot/MartyBugs.net.

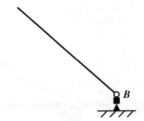

Figure 3.19 An idealization of a portion of the steel bridge structure connected by a very strong *pin* connection (indicated by a circular shape between the line and the square node). The triangular shape suggests that the structure is connected to a pin support at the foundation.

Figure 3.19 is a *portion* of a much larger structure, and the way it appears suggests that the structural member can swing freely like a door hinged at *B*. Such an observation would be true if it were not for the fact that we have not fully completed this idealization. We have not considered any loading in this idealized drawing. As it is drawn in Figure 3.19, we have not included any information about external loading that would have transmitted through the structural members, generating reactionary loading at *B*. Additionally, the structural member is seemingly "cut away" from the rest of the structure. In mathematical terms, that means we have "revealed" internal forces that must be accounted for at some point in our idealized drawing. We will elaborate on these points in limited fashion for the remainder of this chapter.

The reason structural engineers idealize structures is so that we can apply mathematics and physics principles, manifested as static equilibrium, to solve for the magnitude and orientation of internal forces in structural members. For the most part, structures do not *change their shape* when subjected to external loading. Therefore, the application of external loading will generate an *equal and opposite* response of internal forces within the structure itself. The complete structure of a simple Warren *truss* bridge might be idealized, as shown in Figure 3.20.

If external loading were to be applied at the two top joints, then those forces would transmit *through* the structural members in order to reach the foundations at *A* and *B*. An illustration of the flow of these forces is shown in Figure 3.21.

A more exacting analysis of *truss* and *framed* structures will occur in your future engineering mechanics (statics) class. For the purposes of this introductory chapter, we shall focus on how a simple *cantilever* structure might be analyzed using the principles of static equilibrium and the flow of forces.

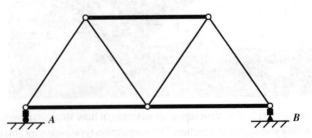

Figure 3.20 A simple Warren truss is made up of structural members arranged as isosceles triangles that are pin-connected at their ends.

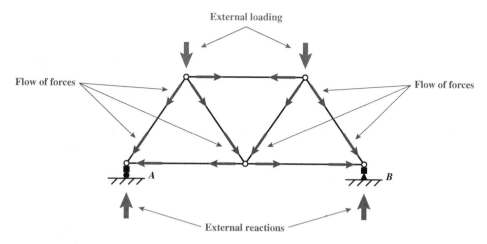

Figure 3.21 External loading (purple) induces internal forces that transmit through structural members and joints to reach the foundation. External reactions manifest that are equal and opposite to the internal forces that reach that particular section of the structure.

A cantilever structure can be oriented horizontally (e.g., cantilever beam) or vertically (e.g., a flagpole). Any external loading can occur in a combination of a concentrated point load at its end and/or a distributed load along its linear length. In order for the cantilever beam to remain rigid and fixed in place, its supporting connection must be a *fixed* node that prevents any translation and prevents any type of rotation (so that cantilever does not swing). In structural engineering, we assume that *three* reactionary forces manifest in order to prevent translation and rotation of a cantilever structure at its fixed-end node (Figure 3.22).

The concentrated point load, P, acting downward at A is at a distance, L, away from the fixed support, B. In order for the cantilever beam to remain where it is, a structural engineer assumes that three reactionary forces *can* manifest and solves their exact values using static equilibrium. Static equilibrium demands that the *sum of the forces in an enclosed system must equal zero*. For a simple cantilever beam, we can apply three independent equations:

$$\sum F_x = 0$$
$$\sum F_y = 0$$
$$\sum M_{\text{any point}} = 0$$

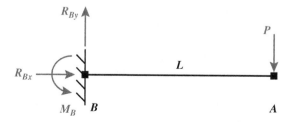

Figure 3.22 Reactionary forces generated at B in response to a concentrated point load of P at A, a distance of L away.

When summing the forces in x, the horizontal axis, we find

$$\sum F_x = R_{Bx} = 0$$

With no other horizontal forces in our system, we can conclude that R_{Bx} indeed equals zero. When summing the forces in y, the vertical axis, we can find

$$\sum F_y = R_{By} - P = 0$$

Here, we must consider that any force pointing in the *positive* y-axis (upward) must be a positive value. We have *guessed* that R_{By} is pointing upward, but we know that P points downward. As such, in our summation of forces, we will treat P as a negative value. When solving for our unknown reactionary force of R_{By}, we can find that

$$R_{By} = P$$

The reactionary force at B is positive and equal to the magnitude of P. Because the result is positive, it suggests that our initial assumption that it points up is valid for our analysis. If the value had resulted in a *negative*, then we would report that the vertical reactionary force at B is pointing downward with a magnitude equal to P.

Finally, we can take a cross product (see Chapter 1) of the concentrated point load, P, and its moment arm, L, to find the generated moment about point B. At point B, there is a reactionary moment, M_B, that is solved for, which resists any rotation at B. To wit,

$$\sum M_B = M_B - L \times P = 0$$

For moment calculations, structural engineers assume that positive moment occurs for counterclockwise rotation. The reactionary moment, M_B, is *assumed* to be positive, so it is drawn in a counterclockwise fashion. The concentrated point load, P, is inducing rotation about point B that is *clockwise*. Therefore, we consider its negative product. We can find, then, that

$$M_B = PL$$

The result for M_B is positive, suggesting that the reactionary moment at B, M_B, is indeed acting counterclockwise based on the external load, P, acting a distance of L away from B.

Let us reconsider the example of the aircraft, particularly its wings. A structural engineer might inspect the idealized drawing of the aircraft (Figure 3.16) and wonder whether the wings are *cantilever* beams because they are connected to a square node, suggesting that no rotation is allowed. To better understand this scenario, a *force system* needs to be *carved out* of the much larger problem in order to inspect the narrower inspection of the aircraft wing.

Force systems are idealized drawings that help structural engineers identify how the external forces manifest specific, internal forces. A structural engineer, for example, might need to understand what the *largest forces are* at a critical juncture point, like where the aircraft wing meets the fuselage. As such, a structural engineer might construct a *force system* to isolate the aircraft wing and treat it as a *cantilever beam* (Figure 3.23). The connection point to the fuselage is now akin to a *fixed-end node*, which allows the structural engineer to approximate the reactionary forces at that location. Those reactionary forces are, instead, the *internal forces* manifesting at the juncture!

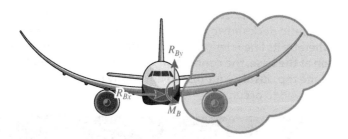

Figure 3.23 A force system constructed around an aircraft wing in order to expose its internal forces at the juncture with the fuselage.

An aircraft wing will be subjected to large external forces when it is midflight, particularly in high-wind conditions. Let us consider a scenario where the weight of the jet engine, W_{engine}, located 25 feet from the fuselage is joined by an upward, uniformly distributed load of ω_{lift}. The constant distribution of this lift force can be solved through an approximation that an *equivalent external load* of $\omega_{lift}L$ acts through the rectangular centroid, 50 feet away from the fuselage.

Using static equilibrium, we can find that the value of the reactions shown in Figure 3.24 is

$$\sum F_x = R_{Bx} = 0$$

$$R_{Bx} = 0$$

$$\sum F_y = R_{By} - W_{engine} + \omega_{lift}(100') = 0$$

$$R_{By} = W_{engine} - 100\omega_{lift}$$

$$\sum M_B = M_B - (25') \times W_{engine} + (50') \times 100\omega_{lift} = 0$$

$$M_B = 25W_{engine} - 500\omega_{lift}$$

Because the values of the jet engine and lift are unknown at this time, we leave our answers in variable form, which enables a structural engineer to explore how the *trade-off* of a heavy engine and lift force influences the maximum forces at critical juncture points. There are many cases where the orientation of the internal forces at the juncture will change in magnitude and change in their orientation.

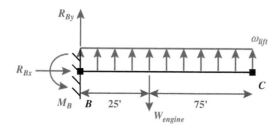

Figure 3.24 A force system depicting an aircraft wing subjected to a concentrated point load of the engine and a uniformly distributed lifting force.

3.16 Consider the vertical post at the center of the truss holding up the roof structure of the patio deck shown in Figure 3.25. Sketch an idealization of the center post whereby its linear length communicates three nodes of interest: the connection at the base, the connection with the bracing supports, and the connection at the top. Justify your placement of any internal hinges where small rotations are allowed to occur.

Figure 3.25   Example of a truss structure for Problem 3.16.
Source: YK/Shutterstock.com.

3.17 Consider the cantilever beam shown. Calculate the reactionary forces at B. Indicate the direction of the reactionary forces (i.e., up, down, left, right, clockwise, or counterclockwise).

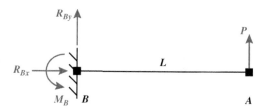

Figure 3.26

3.18 Consider the cantilever beam shown. Calculate the reactionary forces at B. Indicate the direction of the reactionary forces (i.e., up, down, left, right, clockwise, or counterclockwise).

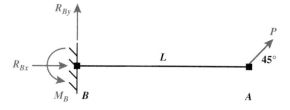

Figure 3.27

3.19 Consider the cantilever beam shown. Calculate the reactionary forces at B. Indicate the direction of the reactionary forces (i.e., up, down, left, right, clockwise, or counterclockwise).

Figure 3.28

3.20 Consider the cantilever beam shown. Calculate the reactionary forces at B. Indicate the direction of the reactionary forces (i.e., up, down, left, right, clockwise, or counterclockwise).

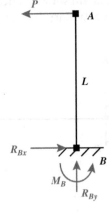

Figure 3.29

3.21 Consider the cantilever beam shown. Calculate the reactionary forces at *B*. Indicate the direction of the reactionary forces (i.e., up, down, left, right, clockwise, or counterclockwise).

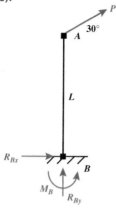

Figure 3.30

3.22 A structural engineer is asking for your help in meeting a client's needs. The client owns a business specializing in dolphin boat tours along the beach. The client would like to erect a very large flagpole outside of their business near the beach that would feature a prominent flag that advertises the boat tours. The client is insisting to use a *very long* pine (Ponderosa) "2 × 4" and would like to have the flagpole be as tall as possible. The structural engineer says that any deflection at the top of the flagpole should be limited to 1/300th of the length of the flagpole (e.g., only one inch of horizontal deflection is allowed for a 300-inch-tall flagpole). Assume that the maximum reactionary force that the pine (Ponderosa) flagpole can sustain at its base is approximately 30,000 lbf. What is the tallest flagpole that you *recommend* for this client's needs? Justify your recommendation.

3.23 The client who owns the dolphin boat tours just realized that they also have a *very long* piece of 2.5-inch-diameter steel rod that they can use for their flagpole. The structural engineer continues to advise that any deflection at the top of the flagpole should be limited to 1/300th of the length of the flagpole. Assume that the maximum reactionary force that the steel flagpole can sustain at its base is approximately 240,000 lbf. What is the tallest flagpole that you *recommend* for this client's needs? Justify your recommendation.

3.24 The client who owns the dolphin boat tours is also in possession of bolts and anchors that can each support 500 lbf of lateral (i.e., shear) force, which occur in the same direction as the horizontal reactionary forces at the base of the flagpole. The structural engineer advises you that the typical arrangement for how these bolts might connect to the concrete foundation is square (four bolts) or circular (six bolts). With this new constraining information in your hands, what is the tallest pine and steel flagpole that you *recommend* for this client's needs? Justify your recommendation.

3.25 The client who owns the dolphin boat tours would like their flagpole to be environmentally sustainable in order to align with their efforts of promoting marine conservation. Based on the carbon dioxide emissions generated per ton of raw material for the two designs you have generated (pine and steel), which do you *recommend* for this client's needs? Justify your recommendation.

3.26 The client who owns the dolphin boat tours would like to move forward with the *more affordable design*. Based on the present-day market prices for the two designs you have generated (pine and steel), which do you *recommend* for this client's needs? Justify your recommendation.

Chapter 4

Sustainable Development Goals:
Engineering for Environmental Sustainability

Objectives

Chapter 4 focuses on environmental engineering and its influence on communities. Students will develop working definitions of environmental sustainability and practice through hands-on design work. Students will also create connections between water resources, fluid dynamics, social systems, water availability, sanitation, and technology.

Student Learning Objectives

- Define sustainability, sustainable development, and environmental engineering.
- Describe how access to water is related to water availability, water treatment, and sanitation.
- List the sustainable development goals and describe how an engineer's work might affect the ability to achieve them.
- Describe how water treatment technologies impact the social, cultural, technical, and economic factors in a community.

4.1 Introduction

Sustainable development in one sense is the desire to improve the worldwide standard of living while considering the effects of economic development on natural resources. As economic and industrial centers continue to develop and transform our landscape, more and more people are looking to these urban centers to improve their standard of living. As a result, current trends show that populations are migrating toward more centralized cities and urban areas. This rural-to-urban migration is putting a significant strain on the regions surrounding these cities and megacities (cities with more than 10 million people). Many countries that are becoming more industrialized are struggling to develop the infrastructure required to provide food, water, sanitation, and shelter for the rural migrants.

In many ways, the narrative that informs our contemporary understanding of sustainable development began over 50 years ago. Scientists, environmentalists, and economists identified several environmental and economic challenges associated with the unprecedented increase in global population growth and natural resource consumption. The United Nations requested that the World Commission on Environment and Development (WCED) formulate "a global agenda for change." The commission articulated the concept of sustainable development in its holistic report *Our Common Future* (WCED 1987, p. 41):

> *Sustainable development is development that meets the needs of the present without compromising the ability of future generations to meet their own needs. It contains within it two key concepts: (1) the concept of "needs," in particular the essential needs of the world's poor, to which overriding priority should be given; (2) the idea of limitations imposed by the state of technology and social organization on the environment's ability to meet present and future needs.*

This definition of sustainable development is often referred to as the ***Brundtland definition***, named after the chairman of the UN commission that produced the report. This definition links three key tenets (known as the three pillars) of sustainability: the environment, society, and the economy. Although the report was published more than a quarter of a century ago, it still serves as an important reference point for discussion of the definition, motivation, and challenges associated with sustainable development.

There are implicit ethical dimensions of the definition of sustainable development that generally build on the concepts of equity, fairness, and justice. It is important for engineers and designers to be able to identify environmental sustainability issues, as they often directly inform the policies and regulations that influence engineering design decisions.

Intergenerational equity refers to the use of natural resources in such a way as to take into consideration the needs of both present and future generations. The UN Rio Declaration in 1992 defined ***intergenerational equity*** more broadly, stating that "the right to development must be fulfilled so as to equitably meet developmental and environmental needs for both present and future generations." Intergenerational equity, then, is the ethical obligation of current societies to consider the welfare of future societies in the context of natural resource use and degradation (Makuch and Pereira 2012).

The Brundtland definition of sustainable development includes a focus on meeting the essential needs of the poor, such as basic human needs for food, water, and shelter. Developing communities often face a systemic lack of access to services that would meet their basic needs. For example, over 2 billion people worldwide lack access to a safely managed source of drinking water, and 3.5 billion lack access to a safely managed sanitation service, with significant disparity within and between countries (United Nations 2022).

As the world population continues to grow, more people require more water and produce more waste products. Engineers, scientists, and policymakers have tried to create a model for development that balances all these considerations. World leaders have focused on the relationship between development, population growth, and natural resource management for many years. One of the most profound statements about these interrelationships is UN *Agenda* 21 (United Nations 1992) from the ***Conference on Environment and Development***, held in Rio de Janeiro:

> *All States and all people shall cooperate in the essential task of eradicating poverty as an indispensable requirement for sustainable development. . . . To achieve sustainable development and a higher quality of life for all people, States should reduce and eliminate unsustainable patterns of production and consumption and promote appropriate demographic policies.*

How can sustainable development be achieved if there are more people using more stuff? To answer this complex question, we must first examine the definition of sustainability and understand the context associated with the term. The term *sustainable* or *sustainability* is widely used in a variety of applications, contexts, and marketing materials. Most of the uses of the word *sustainable* infer a qualitative comparison to something "other." An example would be a fictitious marketing slogan like "Our new green Excellanté automobile is the sustainable solution to yesterday's sport utility vehicle (SUV)." There is little or no quantifiable way to compare the advantages and disadvantages of this product compared to another. Nor is there an attempt to explain how a product can be sustainably produced, sold, and disposed of for any significant period. The marketing of "sustainable" goods and products also typically infers that the "sustainable" product is morally superior to the less sustainable product.

The U.S. Environmental Protection Agency (EPA) defines **sustainability** based on a simple principle: "Everything that we need for our survival and well-being depends, either directly or indirectly, on our natural environment. Sustainability creates and maintains the conditions, under which humans and nature can exist in productive harmony, that permit fulfilling the social, economic, and other requirements of present and future generations" (*Federal Register* 2009). The EPA definition of sustainability implies that all humans and nature have an intrinsic value.

4.1 Describe the concept of sustainability.

4.2 How is "sustainability" different from "sustainable development"?

4.3 What tangible indicators can be measured to demonstrate if sustainability is being achieved?

4.2 Sustainable Development Goals

In 2015, the United Nations adopted the ***Sustainable Development Goals*** (SDGs) as a blueprint for peace and prosperity for people and the planet, now and into the future. The SDGs' primary mission is to end poverty. The developers of the SDGs and the earlier Millennium Development Goals (United Nations 2013) recognized that ending poverty requires advances in health and education, access to water and sanitation, a reduction in inequality, and economic growth while simultaneously preserving the natural resources that provide the requirements for life. Engineers of all types, but particularly civil and environmental engineers, play an important role in developing the infrastructure for meeting the SDGs by 2030.

Many of the SDGs can be addressed only by using the skills and knowledge developed through studying and practicing engineering. Engineers must also cooperate with professionals who have advanced knowledge in economics, education, health care, natural sciences, political science, and social science to develop comprehensive solutions that address the SDGs.

Figure 4.1 The 17 UN SDGs.
Source: Checy/Shutterstock.com.

4.4 List the categories of the SDGs that might require the knowledge and skill set of professional engineers.

4.5 Identify two of the SDGs you listed above that you are most personally interested in and describe why you have an interest in those specific goals.

4.6 Describe the engineering knowledge and skills that might be required to address two of the SDGs listed in the question above.

4.3 Civil and Environmental Engineering

Infrastructure projects such as water and sanitation projects are expensive to implement and require regular revenue to operate and maintain water and sanitation systems. Centralized water and sewer systems have proven to be very effective in improving and protecting water quality and reducing the spread of water-related disease. However, centralized systems require a large capital expense; sufficiently trained personnel to design, build, and operate the systems; and collection of revenue to pay for trained operators and operation and maintenance. Decentralized water and sanitation solutions exist and have also been shown to significantly improve water quality and sanitation. Decentralized systems have a lower capital cost, are generally designed for lower operation and maintenance costs, and as such are often implemented to meet the needs of rural communities. Many developing countries lack the fiscal resources and adequately trained staff required to operate and manage water and sanitation services.

Designing, constructing, operating, monitoring, and maintaining drinking water and sanitation systems is often done by *civil or environmental engineers*. Environmental engineers also help design air quality control systems, environmental resource management strategies, and hazardous and solid waste facilities as well as manage water resource systems. Civil engineers may be responsible for designing broader infrastructure systems, such as water transportation systems, sewer systems, transportation analysis and road development, and built infrastructure, such as bridges, buildings, and utility distribution systems.

The mathematical and scientific analysis of forces and energy is critical for developing the infrastructure systems. Accurate engineering drawings (often called blueprints due to the blue paper required for early large-format printers) are required to communicate these design ideas to those who build these systems. In addition to these fundamental engineering skills, advanced knowledge in biology, chemical processes, differential equations, fluid dynamics, materials, and structural analysis may be necessary to design these complex systems.

Environmental engineers apply the basic principles of science and engineering to design water treatment systems for drinking water and treating human and industrially contaminated wastewater. A proper water treatment design requires knowledge of the constituents of concern, the impact of these constituents, the transformation and fate of the constituents, treatment methods to remove or reduce the toxicity of the constituents, and methods to dispose of or recycle treatment by-products. Civil and environmental engineers often split their time between fieldwork that may require evaluating the conditions at a particular work site and analysis and design work typically done in an engineering office setting.

4.7 Describe how the U.S. Bureau of Labor and Statistics defines a civil engineer, the entry-level education required, and the median salary for civil engineers by reviewing the Field of Degree highlights for students and job seekers in the *Occupational Outlook Handbook*.

4.8 Describe how the U.S. Bureau of Labor and Statistics defines an environmental engineer, the entry-level education required, and the median salary for environmental engineers by reviewing the Field of Degree highlights for students and job seekers in the *Occupational Outlook Handbook*.

4.9 Find job listings for a civil engineering position and an environmental engineering position. Describe the similarities and difference between the education required, experience required, and salary range (if available) for two positions. What other educational degrees or requirements would be acceptable for the job listing you have found? You can find job listings through employment websites, USAJobs.gov, ASCE.org, LinkedIn, and other professional networking websites.

4.4 Centralized Water Treatment Systems

Water quality is determined by measuring factors that are suitable to the potential use. Water pollution issues affect every country in every economic category on Earth; however, the pollutant issues vary with country and region. In most highly developed countries, drinking water is assumed to be free of disease-causing organisms. However, in many developing nations, regular daily access to safe drinking water may be inaccessible in broad areas of the country in both urban and rural settings. When people get sick from drinking polluted water, the cause of the illness is most often due to pathogens, or disease-causing microorganisms present in the drinking water.

In developing countries, much of the drinking water may come from surface waters, such as streams, rivers, and lakes, or from shallow wells if water is available near the surface. Both surface water and shallow wells are easily contaminated by microorganisms from animal and human waste. Deep wells provide access to water that is much more likely to be free of contamination from waterborne pathogens, as shown in Table 4.1, and it may require less sophisticated treatment, as shown in Figure 4.2, than surface water. However, groundwater may contain high levels of dissolved compounds, including arsenic and nitrates, that may be difficult to remove through typical treatment processes. Deep wells may be expensive compared to shallow wells or surface water. Deep wells also require specialized equipment to develop and are therefore found in higher-income communities as well as in rural areas in higher-income countries. The highest-quality water available should be used as the source of drinking water prior to treatment.

Microorganisms are a natural part of the environment; in fact, they outnumber all other species combined on Earth. Microorganisms are necessary to decompose natural and anthropogenic wastes. In their water treatment role, they are very helpful in remediating contaminated water, sediments, and soil in engineered systems. Compared to all types of organisms, a relatively small number cause disease, but these microorganisms are the leading cause of premature death in many low-income regions of the world.

Table 4.1 Fecal coliform levels in untreated domestic water sources in selected countries

Country	Source	Fecal Coliforms/100 mL
Gambia	Open hand-dug wells, 15–18 m deep	Up to 100,000
Indonesia	Canals in central Jakarta	3,100–3,100,000
Lesotho	Streams	5,000
	Unprotected springs	900
	Water holes	860
	Protected springs	200
	Borehole	1
Uganda	Rivers	500–8,000
	Streams	2–1,000

Source: UNICEF (2008), p. 111.

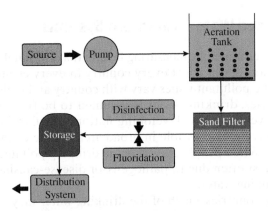

Figure 4.2 Typical water treatment process for potable water use.

Groundwater is usually the preferred source for drinking water (also called potable water) due to the typically low microbial constituent concentration. The water may be pumped from a groundwater aquifer into an elevated storage tank so that gravity provides the force needed for fluid flow through the remaining treatment processes. A gravity flow–based system ensures that some water will be available even in the event of a power outage. The groundwater may be aerated in the storage tank if the water contains elevated levels of dissolved gases, although this is relatively uncommon. The water is typically passed through a granular filtration process. Filtration can improve both the physical and the microbial quality of the water. There are numerous types of granular filters, including slow sand filters, rapid sand filters, and high-rate filters.

Slow sand filters are characterized by a slow rate of filtration and the formation of an active layer called the schmutzdecke. Typically, slow sand filters are completely saturated with water. Large particles are removed in an upper, coarse granular layer. In the lower, finer granular layer, smaller particles are formed, and predatory microorganisms may attack and remove pathogens in the schmutzdecke layer. Slow sand filters are usually operated in parallel because the layers clog over time. Once clogged, the layers must be cleaned or drained, which temporarily upsets the active schmutzdecke layer; time is needed for this layer to re-form to achieve high pathogen removal efficiencies. Slow sand filters may be very effective, as shown in Table 4.2, and are relatively inexpensive and simple to operate and maintain.

The large-scale filtration methods are very successful at reducing pathogen concentrations in water. The filtration methods discussed above, however, may be disrupted by clogging and operational upset, and they do not remove viruses. Disinfection processes are the most effective and reliable method to reduce pathogens in drinking water. Disinfection processes can destroy bacteria, viruses, and amebic cysts, the three types of pathogenic organisms that may be problematic in water. Commonly used chemical disinfection agents used to treat water include free chlorine (OCl^-), combined chlorine ($HOCl$ and OCl^-), ozone (O_3), chlorine dioxide (ClO_2), and ultraviolet irradiation. The effectiveness of the disinfection process is dependent on the pathogen's sensitivity to the disinfectant, the concentration of the disinfectant, the time of contact, and the possible presence of other substances in the water that could interfere with the disinfectant. In many areas, chlorine is the most used disinfectant. It is recommended that the residual chlorine concentration in the water should be 0.5 mg/L, and a minimum contact time of 30 minutes is needed to ensure adequate disinfection.

Table 4.2   Typical removal efficiencies in slow sand filtration

Water Quality Parameter	Effluent or Removal Efficiency	Comments
Turbidity	<1 NTU (nephelometric turbidity unit)	The level of turbidity and the nature and distribution of particles affect the treatment efficiency.
Fecal bacteria	90%–99.9%	Affected by temperature, filtration rate, size, uniformity and depth of sand bed, and cleaning operation.
Fecal viruses and Giardia cysts	99%–99.99%	High removal efficiencies, even directly after cleaning (removal of the schmutzdecke).
Schistosomiasis cercaria	100%	Near complete removal if operation and maintenance are normal.
Color	25%–30%	True color is associated with organic material and humic acids.
Organic carbon	<15%–25%	Total organic carbon.
Trihalomethane precursors	<25%	Precursors of trihalomethanes.
Microcystins	85%–95% or more	Cyanobacteria and their toxins extracted from a cyanobacterial bloom.
Iron, manganese	30%–90%	Iron levels above 1 mg/L reduce filter run length.

Source: UNICEF (2008), p. 103.

4.10 What regulatory agencies are responsible for ensuring that the drinking water on your campus is safe to drink?

4.11 Climate change has severely increased the likelihood that severe floods, hurricanes, and tornadoes will impact the infrastructure even in relatively high-income developed nations. How might these events impact both water supply and sanitation services?

4.12 List the steps in a disaster response plan and describe the role an experienced engineer might play in responding to a natural disaster.

4.13 Hurricane Maria is the worst natural disaster on record to affect Dominica and Puerto Rico, and it is also the deadliest Atlantic hurricane since Hurricane Jeanne in 2004. Hurricane Maria caused catastrophic damage and numerous fatalities across the northeastern Caribbean. Total losses from the hurricane were estimated at upwards of $91.61 billion (2017 U.S. dollars), mostly in Puerto Rico, ranking it as the third-costliest tropical cyclone on record. Hurricane Maria forced many water and sanitation systems offline. How would someone determine if drinking water is safe after a possible system failure? Explain how lack of power and lack of water may have contributed to the high mortality rate associated with the hurricane.

4.5   Decentralized Water Treatment

Water quality is influenced by natural and human factors. Access to safe drinking water and improved sanitation is the theme of SDG number 6. However, access to water and sanitation directly impacts many of the SDGs. Declining water quality is a concern due to the stresses imposed by increased human water consumption, agricultural and industrial discharges, and many negative effects related to our rapidly changing climate. Poor water quality is a concern due to the economic, environmental, and social impacts that occur if water supplies become too polluted for drinking, washing, fishing, or recreation.

The UN SDG progress report (United Nations 2022) states, "Access to safe water, sanitation and hygiene is the most basic human need for health and wellbeing. Billions of people will lack access to these basic services in 2030 unless progress quadruples. Demand for water is rising due to rapid population growth, urbanization and increasing water needs from agriculture, industry, and energy sectors. Decades of misuse, poor management, over-extraction of groundwater and contamination of freshwater supplies have exacerbated water stress. In addition, countries are facing growing challenges linked to degraded water-related ecosystems, water scarcity caused by climate change, underinvestment in water and sanitation, and insufficient cooperation on transboundary waters."

Experts in the field of sustainable development believe that providing clean water and sanitation for sub-Saharan Africa is one of the world's greatest challenges. The United Nations reports that approximately 70% of the population in sub-Saharan Africa lacks access to improved water sources and that 79% of the population lacks access to basic sanitation facilities. Centralized water treatment is not a feasible option for many communities' drinking water supplies because it is extremely expensive to construct and maintain the required infrastructure.

In much of sub-Saharan Africa, the primary ways in which people obtain clean drinking water are by boiling water or purchasing imported bottled water. Boiling the water requires wood and native vegetation, depleting local resources and emitting smoke into households and the atmosphere. Buying imported water is not a cost-effective, long-term solution for low-income populations. As a result, many people in Africa drink water that does not meet the drinking water standards set by the World Health Organization (WHO). Providing the technology to implement point-source water treatment in the community may decrease childhood and maternal mortality rates in many communities in Africa.

The semiarid and arid areas of Kenya are a perfect case study of the typical use of water in most of sub-Saharan Africa. Kenyan communities are highly influenced by the availability and use of limited water sources. The impact this has on communities is far greater than the traditional public health metrics. Gilbert, an engineer from Northern Kenya, shared his personal experiences of what it was like to live without access to an adequate water supply.

Gilbert was born in the small community of Arpollo nestled against the eastern base of the Cherangani range in the Rift Valley. In Arpollo, one experiences the rawness of nature's basic call for water and the stark fact that the water must be carried to the community from a river five miles to the east out on the Rift Valley floor. This Rift Valley river water is subject to all the uncertainties found in any Kenyan river, which hosts the entire range of contaminant sources afflicting human health.

Due to rural-to-urban migration, Gilbert's parents moved and settled in Kapenguria, situated 20 miles northeast of Kitale. Kapenguria had a small municipal water system that is over 20 years old. The district department of the Ministry of

Water and Irrigation oversaw operation and maintenance of the water system. There were two water tanks in the system. The primary tank was located 15 yards from the river, the source of the water. Water from the river was lifted into the first tank by an electric pump. Basic sediment removal and screening was performed before it was pumped up to the secondary tank. The secondary tank was situated on the ground almost a kilometer from the primary tank on a slope approximately 30 meters higher than the primary tank. In the secondary tank, the water was chlorinated before being distributed to homes, restaurants, clinics, and the hospital.

Since commissioning the water system, maintenance of the system was sporadic. Over time, water access diminished due to lack of ability to collect customer fees, illegal connections to the distribution lines, and corruption. Revenue was insufficient to repair primary infrastructure problems, such as broken distribution lines and broken electric pumps. This water system eventually delivered water to less than 20% of its original customers.

Complaints to the relevant authority went unanswered, which prompted Gilbert's family to look for other sources of water. Gilbert's family had to fetch water from the local river after the pipes connecting their home to the distribution system failed. In Gilbert's family and many others throughout the world, fetching water from the river remains unsafe, time consuming, and unreliable.

4.14 What percentage of the population of Kenya has access to water in (a) urban areas and (b) rural areas according to the UN report on SDGs (www.sdg6data.org/en/snapshots)? (c) Is access to clean water expected to get better or worse in the coming years?

4.15 Create a digital story or poster that describes the SDGs that are related to Gilbert's experience in Kenya. You may use one of the following technologies to help develop your submission: Canva, Piktochart, PowToon, Prezi, or Weebly.

Gilbert's family's experience with water systems is common in many communities throughout the world. Historically, there is a high failure rate of water devices in developing countries, with 30% to 60% critical failure rates of existing water supply devices (Henriques and Louis 2011). Local communities are often not trained to repair the water systems. Consequently, about 35% of the installed systems fail (Henriques and Louis 2011). Many communities, like the one in Kenya that Gilbert described, benefit from low-cost, low-maintenance treatment systems that can be used in the home, called ***point-of-use*** (POU) or ***household water treatment*** (HWT) technologies.

POU treatment devices are based on the same principles used in larger-scale centralized treatment plants. There are many types of POU devices. Typically, POU devices utilize granular media (sand) filtration, ceramic filtration, chlorine disinfection, or ultraviolet (UV) disinfection. POU devices may be added to water plumbing in the home if it exists or to units that require water to be collected and treated separately by the POU device. Water for cooking or drinking should come only from the POU device.

An effective POU device must be portable, easy to use, and able to treat the average household use of drinking water: 80 L of water per day (Striebig et al., 2010). The output of treated water that comes from the device must be equal to or greater than the total water consumption per family per day. The water must also be free from harmful treatment by-products, and the water's appearance and taste must be acceptable to the user for the technology to be implemented in homes. Perhaps most important, the initial cost and operational cost of the POU device must be affordable for the community.

Three treatment technologies (shown in Table 4.3) with a proven and technologically appropriate track record may be considered for drinking water purification. The advantages and disadvantages of each technology should be evaluated. Each technology fits a specific niche in developing countries. The WHO (2022) has developed an International Scheme to Evaluate Household Water Treatment Technologies ("Scheme") to evaluate the performance of HWT technologies against WHO microbial health-based criteria. The results of the Scheme evaluation are intended to help communities and organizations choose an appropriate water filtration method. These different treatment methods vary in their ability to destroy pathogens that pose a serious health risk: bacteria, viruses, and protozoa. These recommendations provide the basis for evaluating and classifying HWT into three levels of performance—3-star (★ ★ ★), 2-star (★ ★), and 1-star (★)—based on their ability to remove pathogens from drinking-water (www.who.int/tools/international-scheme-to-evaluate-household-water-treatment-technologies).

Table 4.3   Examples of POU water treatment devices

Technology	Advantage	Disadvantage	Filter Time
Biosand™ sand filtration	High removal efficiency for microorganisms	Needs continual use and regular maintenance Cost	1–2 hours
Filtrón™ ceramic filter	High removal efficiency for microorganisms Sized for households Relatively inexpensive	Requires fuel for construction Limited lifetime Requires regular cleaning	2–8 hours
SODIS™ solar water disinfection	Highly effective Inexpensive Can reuse a waste product (PET bottles)	Long treatment time (12 to 48 hours) Does not remove other pollutants Requires warm climate and sunlight	12–48 hours

Source: Based on Striebig et al. (2007). "Activated Carbon Amended Ceramic Drinking Water Filters for Benin." *Journal of Engineering for Sustainable Development* 2, no. 1:1–12.

Sustainable Development Goals: Engineering for Environmental Sustainability

Cost Appropriateness of POU/HWT

4.16 One example of a POU device that is simple to employ is a product called the AquaTab, which uses a solid chlorine-forming disinfectant process. In 2022, the cost for 100 tablets that each treat four gallons of water was $24.95 (U.S. dollars). The AquaTabs are simple to use in sub-Saharan Africa because the rural communities have already been exposed to the tablets before and the instructions for using the tablets are a part of the packaging. How much would it cost to use AquaTabs for one year as a POU treatment in a home that requires 80 L of water per day?

4.17 The average annual wage in Kenya in 2022 was approximately $1,524. What percent of the household income must be spent on AquaTabs if there is one wage earner in the home? Compare this to how much you are accustomed to paying for water. (You may have to ask the person in your household who pays the water bill how much that expense costs per month and what the average household income is each year.) Does it seem like a reasonable expectation to expect to pay this percentage of the household income for water treatment in a Kenyan community that does not have a reliable water supply?

4.18 Choose an alternative HWT option from the WHO Scheme. Use the information from the WHO evaluation (www.who.int/tools/international -scheme-to-evaluate-household-water-treatment-technologies) and information you find online to compare an alternative process to the AquaTab water treatment process. You might consider cost, effectiveness, and other criteria in your evaluation.

The *biosand water filter*, illustrated in Figure 4.3, is an adaptation of slow sand filtration that is designed for use by families at the household level. This water filtration technology was developed by Dr. David Manz, a former University of Calgary professor. Biosand filters have been implemented throughout the world and have been demonstrated to successfully reduce water-related disease.

The biosand filter is like a traditional slow sand filter; however, it is smaller and does not require constant water flow. This POU treatment device is made from a concrete or plastic container, a diffusion plate that is placed above layers

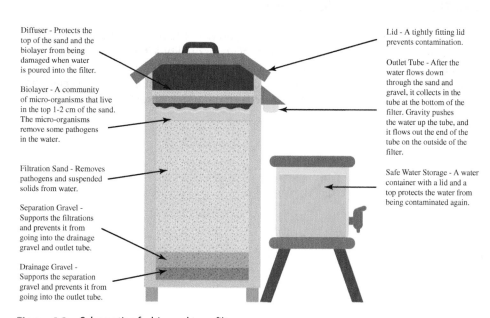

Diffuser - Protects the top of the sand and the biolayer from being damaged when water is poured into the filter.

Biolayer - A community of micro-organisms that live in the top 1-2 cm of the sand. The micro-organisms remove some pathogens in the water.

Filtration Sand - Removes pathogens and suspended solids from water.

Separation Gravel - Supports the filtrations and prevents it from going into the drainage gravel and outlet tube.

Drainage Gravel - Supports the separation gravel and prevents it from going into the outlet tube.

Lid - A tightly fitting lid prevents contamination.

Outlet Tube - After the water flows down through the sand and gravel, it collects in the tube at the bottom of the filter. Gravity pushes the water up the tube, and it flows out the end of the tube on the outside of the filter.

Safe Water Storage - A water container with a lid and a top protects the water from being contaminated again.

Figure 4.3   Schematic of a biosand type filter.

of sand and gravel inside, and a carefully constructed outlet to ensure that water added to the top of the filter flows through the entire filter bed. Microorganisms and suspended material are removed through the same processes that occur in a larger-scale sand filter described above.

Pure Water for the World, Inc. (PWW) is a nonprofit NGO headquartered in the United States and operating out of Honduras and Haiti. It is dedicated to improving lives by empowering people with access to life's most basic necessities: safe water and sanitation. PWW started implementing concrete biosand filters (BSFs) in Honduras in the early 2000's and in Haiti in 2008. Later, it switched to Hydraid® plastic filter containers to keep the price of the system low and make it easier to transport the filters. To date, PWW has installed BSFs in more than 16,975 homes, reaching 100,000+ people, and in more than 2,080 schools, reaching 500,000+ students. Through financial support from foundations, grants, and donations, PWW is able to subsidize the filter costs; however all families are required to contribute a small portion of the cost (<$10 USD) to ensure a strong sense of ownership.

PWW has invested in training community agents (CAs), members of the local communities who volunteer to serve as an extension of the PWW team, helping to install the filters and support families with the correct and consistent use of the filters and hygiene practices. CAs also provide feedback to PWW and do the follow-up visits scheduled for each household. Interested community members are eligible to become CAs, and women are encouraged to participate. The CAs monitor the biosand filters and go with a PWW staff health promoter to visit each filter. Together they check that the filter was installed properly, reinforce the proper use and mainte-nance of the filter, and solve any question the users have. PWW works to build and expand local capacity through comprehensive WASH (water, sanitation, and hygiene) programs, including the BSF technology and CA programs. PWW also presents work-shops to raise awareness and build capacity among the communities about general WASH issues, including environmental hygiene, household hygiene, latrines, and per-sonal hygiene.

Building Your Own Water Filter

4.19 Consider what you know about how you would like your water to look, smell, and taste, as well as other characteristics of water quality. What characteristics are required for implementation of a sustainable POU water treatment system in a community in a low-income country? Consider the technical, economic, social, and environmental factors that contribute to your definition of sustainable.

4.20 Estimate the amount of water you consume each day from your water bill or from all your water-related activities, such as drinking, washing dishes, washing clothes, bathing/showering, cleaning, and so on. You answer should have units of liters of water per day per person. Use mathematical calculations to show how you made your estimate. What type of POU/HWT device would meet your needs for water?

Table 4.4 Materials list for a bucket sand filter

Material	Amount	Cost ($ U.S.)	Amount Per Filter	Cost Per Filter
Sand	0.5 cu ft	3.38		
Gravel	0.5 cu ft	3.48		
Bucket	1	2.54		
White spigot	1	1.76		
Black spigot with gaskets	1	2.81		
PVC pipe	10 ft	1.68		
PVC cement	1 pack	7.51		
Epoxy	1 pack	5.98		
Male adapter	1	0.23		
Elbow	1	0.26		
Threaded elbow	1	0.57		
End cap	1	0.32		
O rings	10 pack	1.97		
Other				

4.21 Based on the materials list in Table 4.4 and the biosand filter schematic, create engineering drawings that can be used to fabricate a sand filter with the materials listed. List each part and illustrate how the parts will fit together.

4.22 Using Excel or a similar spreadsheet, calculate how much your filter materials cost. What parts are most expensive? Where might you be able to possibly achieve the greatest reduction in cost?

4.23 Construct your sand filter based on your design and engineering drawing using the tools and materials provided. Include a photo of your as-built sand filter.

4.24 Collect or use the tap water provided and test this water as a control test using the PathoScreen methodology (www.hach.com/asset-get.download .jsa?id=7639984011). Report your results.

4.25 Test the integrity of your filter by running tap water through your filter at a slow rate to identify any possible leaks in your filter and to clean the sand in the filter. Test the water exiting the water filter using the PathoScreen methodology. Report your results.

4.26 Walk to the nearest surface water source with a jerrican provided and collect water or use the untreated surface water provided. Test this water prior to filtering using the PathoScreen methodology. Report your results.

4.27 Collect or use the untreated surface water provided. Pour at least 8 L of the water through the filter. Time how long is required to obtain 8 L of filtered water. Discard the first 7 L of water collected from the exit of your sand filter. Collect the eighth liter of water and test this water exiting the filter using the PathoScreen methodology. Report your results in a written format and with a photo of the test tubes.

4.28 Describe the effectiveness of your filter design. Include in your analysis the total cost, average water filtration rate per hour, and the effectiveness of treatment based on the PathoScreen methodology.

4.29 Would your simple filter design be appropriate and beneficial for use in a low-income country community? Explain your response based on technical, economic, social, and environmental factors.

4.30 Describe the requirements to operate and maintain the filter in the home. What training and materials would be needed to ensure that the system worked effectively for a period of five years?

References

Federal Register. (2009). Executive Order 13514. "Federal Leadership in Environmental, Energy, and Economic Performance onto the Agency's Green Purchasing Plan." Washington, DC. 74(194): 52117–52127.

Henriques, J., and Louis, G. (2011). "A Decision Model for Selecting Sustainable Drinking Water Supply and Greywater Reuse Systems for Developing Communities with a Case Study in Cimahi, Indonesia." *Journal of Environmental Management* 92: 214–222.

Makuch, K. E., and Pereira, R. (2012). *Environmental and Energy Law*. Chichester: John Wiley & Sons.

Striebig, B., Atwood, S., Johnson, B., Lemkau, B., Shamrell, J., Spuler, P., Stanek, K., Vernon, A., and Young, J. (2007). "Activated Carbon Amended Ceramic Drinking Water Filters for Benin." *Journal of Engineering for Sustainable Development* 2, no. 1:1–12.

Striebig, B., Gieber, T., Cohen, B., Norwood, S., Godfrey, N. and Elliot, W. (2010). "Implementation of a Ceramic Water Filter Manufacturing Process at the Songhai Center: A Case Study in Benin." Paper presented at the International Water Association World Water Congress, Montreal.

UNICEF. (2008). *UNICEF Handbook on Water Quality*. New York: UNICEF.

United Nations. (1992). *Agenda 21*. Paper presented at the United Nations Conference on Environment and Development. Rio de Janeiro.

United Nations. (2013). *The Millennium Development Goals Report 2012*. New York: United Nations.

United Nations. (2021). *Progress towards the Sustainable Development Goals Report of the Secretary-General*. United Nations Economic and Social Council. E/2021/58.

World Commission on Environment and Development. (1987). *Our Common Future*. New York: Oxford University Press.

World Health Organization. (2022). *International Scheme to Evaluate Household Water Treatment Technologies*. www.who.int/tools/international-scheme-to-evaluate-household-water-treatment-technologies.

Chapter 5

Food, Water, and Nutrients in Chesapeake Bay: An Earth Systems Engineering Approach

Objectives

Chapter 5 focuses on the concepts of systems thinking and analysis for complex engineered systems. Students will develop basic knowledge and tools to identify a system, decompose it into parts, define interactions, perform analysis, and apply control measures if necessary. Students will also create connections between human systems, economic systems, and ecological systems.

Student Learning Objectives

- Apply the basic concepts associated with systems analysis.
- Define the input flows, output flows, and interactions between different parts of a system model.
- Develop system representations at multiple levels of fidelity.
- Evaluate trade-offs to make informed decisions.
- Synthesize systems thinking, principles, and tools to complex systems.

Examples of ecological systems might include analyses of trout populations in headwater streams, impacts from agriculture and residential development in piedmont regions of the watershed, impacts of recreational boating, and impacts to fish and oyster populations in Chesapeake Bay.

5.1 Introduction

The interactions between human-made infrastructure and nature systems are complex and changing at the greatest rate in human history. These interactions significantly impact human development, human infrastructure, and ecological systems (Allenby 2002, 2007; NASA 2003). Engineering infrastructure must be more resilient due to the changing climate and future resource limitations. Systems engineering concepts can be introduced to illustrate how these systems interact (Gorman et al., 2003; Gorman 2005). The Chesapeake Bay watershed is complex and extremely large and consists of multivariable systems that are difficult to define and model using simple single-variable engineering analysis.

   Scientists and engineers can evaluate global systems today using satellites and computational models to better understand how changes occur to the Earth's global-scale system, referred to as *global change*. Systems that are changing globally include planetary-scale changes to atmospheric circulation; ocean circulation; climate; carbon, nitrogen, water, and other cycles; sea-ice changes; sea-level changes; food webs; biological diversity; pollution; health; fish stocks; and more. Both local and global changes have influenced Chesapeake Bay, North America's largest estuary (see Figure 5.1). An *estuary* is a place that is influenced by fresh water from river systems as well as salt water from ocean tides. Estuaries are some of the most

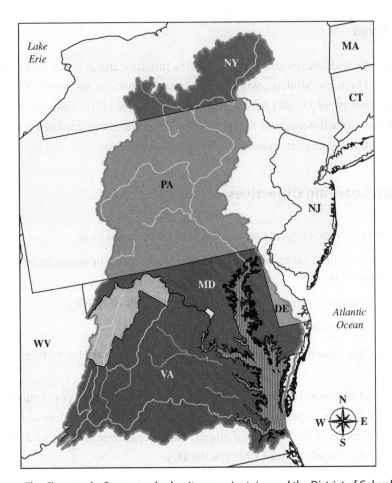

Figure 5.1   The Chesapeake Bay watershed as it spans six states and the District of Columbia.
Source: https://www.nrcs.usda.gov/wps/portal/nrcs/detailfull/national/programs/initiatives
/?cid=stelprdb1047323.

biologically diverse and productive areas on the planet. However, Chesapeake Bay's productivity and diversity have been decreasing for the past 100 years due to human influences that have altered land use, water volume, water composition, and the climate. No single factor is responsible for the decline of Chesapeake Bay's biological functions; instead, there are complex interactions that influence how the Chesapeake Bay estuary has evolved in response to human-induced change. To understand and improve the biological functions on the bay, we must learn how to analyze multivariable complex systems.

5.2   Earth Systems Engineering

The term "Earth system" refers to Earth's interacting physical, chemical, and biological processes. The system consists of the land, water, atmosphere, and icy poles. It includes the planet's biogeochemical cycles, such as the carbon, water, nitrogen, and phosphorus cycles, as well as deep Earth processes. We are now living in what many scientists believe is a new age on the planet call the Anthropocene. In this new age, the human influence on our planet has greatly changed the naturally occurring processes that have occurred for the past thousands of years. Human's social and economic systems are now embedded within the Earth's systems. Human systems have altered the Earth system by changing waterways, changing land cover, and changing the planet's climate cycles. Earth systems engineering recognizes and seeks to understand the human influences on the Earth's systems, examples of which are illustrated by NASA in Figure 5.2, and considers ways to manage human impacts to minimize the detrimental effects of many of these changes:

> *This requires an identification and description of how the Earth system is changing, the ability to identify and measure the primary forcings on the Earth system from both natural and human activities, knowledge of how the Earth system responds to changes in these forcings, identification of the consequences of these changes for human civilization, and finally, the ability to accurately predict future changes with sufficient advanced notice to mitigate the predicted effects.* (Dr. Blanche Meeson, assistant director of Earth sciences for education and outreach, NASA, Goddard Space Flight Center, May 2000)

Earth system science builds on the fundamental disciplines of biology, chemistry, computer science, engineering analysis, math, and physics, in addition to the social sciences, which help us understand and predict human behavior. The Earth system is complex due to the huge number of interacting components. The interactions between the biological, human, and physical components are difficult to predict. This means that these interactions are difficult to model with mathematical precision in part due to the inherent difficulties in anticipating human behavior and the response

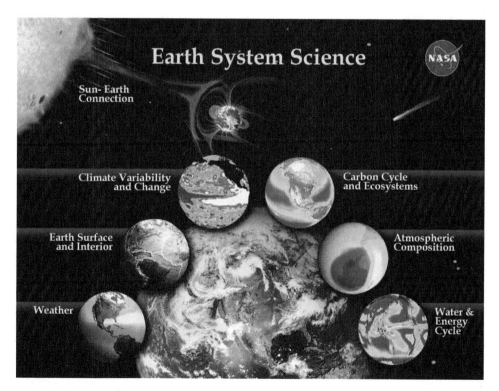

Figure 5.2 NASA's depiction of Earth systems.
Source: NASA.

of Earth systems to unpredictable human behavior. Traditional single-variable and top-down approaches used to solve relatively simple problems are inadequate to address these huge multivariable and interdependent issues in large-scale ecosystems like Chesapeake Bay. Problem solvers must adopt approaches that can interpret the complexities and inherent uncertainties of the complex Earth system. Implementing an Earth systems solution to address issues like the decline of biological productivity in Chesapeake Bay must include basic research to understand interactions and feedbacks, sustained long-term monitoring and observation to verify basic research, and regular evaluation of progress to adapt to unexpected changes in the system.

5.1 Describe how the *INCOSE, the International Council on Systems Engineering*, defines a systems engineer.

5.2 Find job listings for a *systems engineering* position. List the duties or the description of job tasks the employee is expected to do for three advertised positions. You can find job listing through employment websites, USAJobs .gov, INCOSE.org, LinkedIn, Indeed.com, and other professional networking websites.

5.3 List four physical processes and four biological processes that *you* interact with each day.

5.4 Create a box to represent two of the physical and two of the biological processes listed above that *you* interact with each day. Add to each box a process name, location, interactions between other processes, and the estimated quantity of material or energy involved in the process.

5.5 Draw a schematic of the *"you"* system that shows how materials and energy flow around you in your daily life.

5.3 The Blueprint for Chesapeake Bay

The Chesapeake Bay is the largest estuary in the world. The 11,684 miles of shore-line is greater than that of the entire western coast of the United States. Significant efforts have been taken to restore the water quality and living resources of the bay throughout the 64,000-square-mile Chesapeake Bay watershed. Yet despite these efforts, much of the water, habitat, and fisheries in the bay remain degraded and jeopardized. The Chesapeake Bay Foundation (CBF) measures the current state of the bay compared to the Chesapeake historically described by Captain John Smith in his exploration narratives from the early 1600s. The CBF issues its *State of the Bay Report* based on 13 indicators of pollution, habitat, and the health of fisheries in the bay. In 2020, the score declined one point to 32, a D+ overall, as illustrated in Table 5.1.

There are many system-wide challenges to improving Chesapeake Bay. The *State of the Bay Report* uses 13 separate indicators, or measurements, to describe the overall health of Chesapeake Bay. However, while each indicator can be mea-sured separately, all the indicators are interrelated in the system. For example, the number of fish and animals in the bay are dependent on the habitat and the water quality. The type of animal life present or absent influences the plants, structures, soils, and sediment within the bay that contribute to the available habitat quality. Better riparian buffers and more abundant seagrass and oysters directly influence the water quality.

Over the past two decades, most indicators of Chesapeake Bay health have improved somewhat, as shown in Table 5.1. However, this improvement has been very slow, and these metrics indicate a system still far from its ecological potential. The polluted water, land changes, and loss of bioproductivity have negative effects on the ecosystem, regional economy, and outlook for the communities surrounding the bay.

Table 5.1 The CBF's *State of the Bay Report* for 2020 showed improvements in many measured water quality indicators, but much work still needs to be done to improve habitat and fisheries in Chesapeake Bay.

	Indicator	2000	2020	Change from 2000 to 2020	"Grade" in 2020
Pollution	Nitrogen		17		F
	Phosphorus	15	27	12	D
	Dissolved oxygen	15	44	29	C
	Water clarity	15	17	2	F
	Toxics	30	28	−2	D
Habitat	Forest buffers	53	56	3	B
	Wetlands	42	42	0	C
	Underwater grasses	12	22	10	D−
	Resource lands	33	33	0	D+
Fisheries	Rockfish	75	49	−26	C+
	Oysters	2	12	10	F
	Blue crabs	46	60	14	B+
	Shad	5	7	2	F

Source: Chesapeake Bay Foundation. *State of the Bay Report* (2000, 2020).

Identifying Potential Problems in a System

Identify applications of engineering sciences for evaluation of indicators or other issues related to Chesapeake Bay. Use your personal experiences to identify a problem of personal interest:

- Think critically about how you can apply engineering science to
  - identify a problem statement,
  - create a mathematical representation of the problem, and
  - solve the problem using mathematical and engineering relationships.
- Use the scientific method (hypothesis-driven research) to refine your curiosity about the interconnected Earth system in which we all live.

5.6 What questions do you have about how the Chesapeake Bay system functions in relation to the water quality, land use, fisheries, and people interacting with the bay environment?

5.7 What *one* question would you most like to explore and learn more about regarding the Chesapeake Bay system?

5.8 Describe what steps you can take to explore your curiosity. How would you find out more information related to Problem 5.7?

5.9 List the potential *sources* of information that might help you learn more about your above question (see Problem 5.7). Also, list where you might find this information (e.g., the library, a specific website, an educational organization).

5.10 What are some characteristics of "good" or "reliable" information? How would you interpret and prioritize various sources of information from more reliable to less reliable?

5.11 List a comprehensive set of key words related to your area of interest. You will later use these key words to search sources for information that helps describe how the Chesapeake Bay system functions.

5.12 In this step, draw a picture or sketch that represents the system you are curious about. Sketch processes within the system and define boundaries for a system that encompasses your idea. Later, you will refine your sketch, so this sketch should be done by brainstorming and visually representing your ideas and things you think might be connected to your topic of interest.

5.4 Modeling Processes

Mathematical models are helpful to predict how changes to a system may affect part of the system or the overall system. For example, if farmers in the Chesapeake Bay system were to change the amount of nitrogen- or phosphorus-based fertilizer they use in the Chesapeake Bay watershed, that would likely affect the water quality measurements in the bay. We measure the amount of material by mass. The *law of conservation of mass* states that mass cannot be created or destroyed. Balancing the mass and energy flow into and out of a system allows engineers and scientists to quantitatively analyze the behavior of processes and systems. Engineers create *black box models* that are simple representations of how mass and energy flows through a defined system, as shown in Figure 5.3.

Imagine a box with material flowing through it, as shown in Figure 5.3. The flows into the box are called *influents*, represented here by X. If the flow is described as a mass per unit time, X_0 is the mass per unit time flowing into the box. Similarly, X_1 is the outflow, or *effluent*. In this case, no material is being created or destroyed within the box, and the flow remains constant with time, that is, at steady-state flow. The material balance using the box as the *control volume* of our analysis yields the following:

Mass per unit time of X IN		Mass per unit time of X OUT

$$[X_0] = [X_1] \qquad (5.1)$$

The black box can be used to establish a model for flows into and out of a process. When we combine processes together, we can replicate more complicated systems.

When flow is contained in pipes or channels, it is convenient to place the control volume around any junction point. If the flow is not contained, it may be approximated by visualizing the system as a pipe flow network. Rainwater falling to Earth, for example, can either percolate into the ground or run off into a stream or river. This system can be visualized, as shown in Figure 5.4, and a material balance may be performed on the imaginary black box.

The amount of material or energy flow into and out of a system is also defined by the boundaries of the system under examination. The volume within these boundaries defines the control volume. A control volume in the chemical reactor in a chemistry laboratory is defined by the boundaries of the reactor's walls. A watershed is often used as a boundary in modeling Earth's systems since it is the boundary between

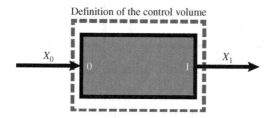

Figure 5.3   A black box process with one inflow and one outflow.

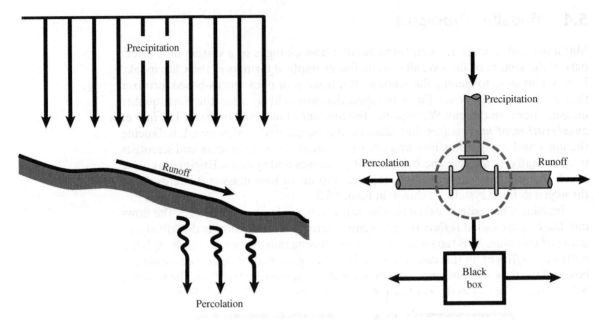

Figure 5.4 Precipitation, runoff, and percolation can be visualized and modeled as a black box.

two land-drainage areas. A watershed is the land area in which rainfall falling onto that land drains into a common outflow. Thus, the control volume for surface water flowing into Chesapeake Bay is defined by the boundaries of the watershed. Surface water flowing out of Chesapeake Bay exits into the Atlantic Ocean.

The general procedure for solving material balance problems involves the following:

1. Drawing the system as a diagram, including all flows (inputs and outputs) as arrows.
2. Adding all available information provided, such as flow rates and concentrations, and assigning symbols to unknown variables.
3. Drawing a continuous dashed line to represent the control volume around the components or components that are to be balanced. This may include a unit operation, a junction, or a combination of these processes.
4. Deciding what material is to be balanced. This may be a volumetric or mass flow rate.
5. Writing the general material balance equation:

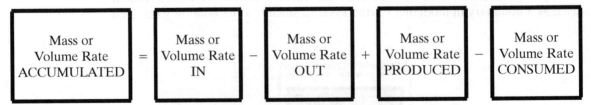

6. Solving for the unknown variables using algebraic techniques.

A flow into a box may be separated into two or more effluent streams, as shown in Figure 5.5. The flow into this box is X_0, and the two flows out are X_1 and X_2. Steady-state conditions in a process assume that no material is being created or destroyed within the box, and the flows shown in Figure 5.5 remain constant with

Definition of the control volume

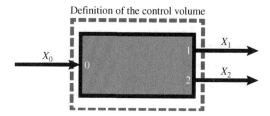

Figure 5.5 A process with one influent and two effluents.

time. An example of this can be seen in Figure 5.6 in the braiding of channels in an estuary. The main channel of the stream carries sand to and from the oceans. The sand deposits in various places, causing smaller channels to split and divide. The material balance for the splitting system is

| Mass per unit time of X IN | = | Mass per unit time of X OUT |

$$[X_0] = [X_1] + [X_2] \tag{5.2}$$

The material X can be separated into more than two fractions. The material balance for n effluent streams becomes

$$[X_0] = \sum_{i=1}^{n} [X_i] \tag{5.3}$$

A black box can also receive numerous inflows and mix the flow streams together to discharge one effluent, as shown in Figure 5.7. The flows into this box

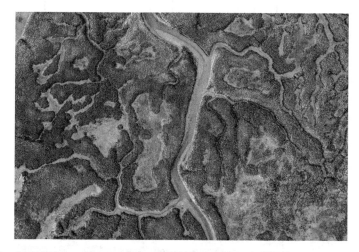

Figure 5.6 An aerial view of an estuary of San Francisco Bay in Marin County, California, that shows the main river channel splitting into multiple channels. As the flow in an estuary goes out at low tide and flows inland at high tide, the multiple flow directions result in changing deposits of sand and sediments, causing the braided channel in the estuary. Braiding, or flow splitting, may also occur in inland rivers and streams.
Source: DianeBentleyRaymond/E+/Getty Images.

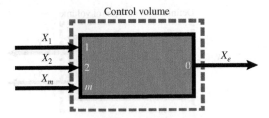

Figure 5.7 A mixer with several (*m*) inflows and one effluent flow.

are X_1, X_2 ..., X_m. The material balance for the mixing system yields the reverse of the splitting system:

$$\sum_{i=1}^{m} [X_i] = [X_e] \qquad (5.4)$$

Example 5.1   Runoff from Fields into a Stream

Wheat requires fertilizer as a commercial agricultural crop. As the wheat plant grows, it consumes nitrogen, as illustrated in Figure 5.8. A farm located in the Chesapeake Bay watershed applies 128 kg of nitrogen fertilizer to one hectare (10,000 m²) of a wheat field over the course of a year. The wheat utilizes 74 kg of nitrogen as it grows. Microorganisms in the soil consume 6 kg of nitrogen over the year. During this year, 0.54 m of rain falls on the land. Sixty-one percent of the rainfall is used by the crops and evaporates into the air. Twenty-two percent of the rainfall flows over the field and into a nearby ephemeral stream as runoff. A black box model of the nitrogen utilization process is shown in Figure 5.9.

a) What is the mass of nitrogen per year that is removed from the soil by runoff?

b) What is the concentration of the nitrogen in the runoff in units of mg/L?

Set up the mass balance for nitrogen utilization in terms of kilograms per year:

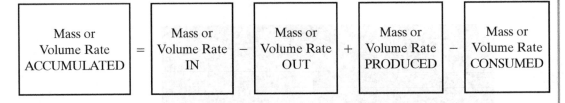

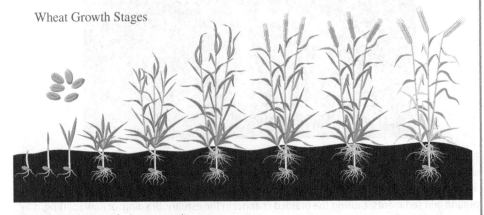

Wheat Growth Stages

Figure 5.8   Stages of wheat growth.
Source: Andrii Bezvershenko/Shutterstock.com.

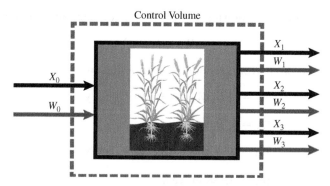

Figure 5.9 A black box process model of nitrogen utilization in a wheat field.
Source: Andrii Bezvershenko/Shutterstock.com.

Notice there are two input streams and two output streams for our control volume around the one-hectare farm field. We will define X as the mass flow of nitrogen and W as the water flow. Defining each variable for nitrogen within the control volume yields

X_0 = Nitrogen mass added to the one-hectare wheat field = 128 kg/year

X_1 = Nitrogen mass removed by wheat in the one-hectare wheat field = 74 kg/year

X_2 = Nitrogen mass removed by microorganisms in the one-hectare wheat field = 6 kg/year

X_3 = Nitrogen mass removed by water from runoff from the one-hectare wheat field = ?

> | Mass per unit time of X IN | = | Mass per unit time of X OUT |

Part a:

[Mass per unit of nitrogen IN] = [Mass per unit of nitrogen OUT]

$$X_0 = X_1 + X_2 + X_3$$

128 kg/year = 74 kg/year + 6 kg/year + X_3 kg/year

Solving for the unknown yields X_3 = 48 kg/year

Part b:

Defining each variable for water within the control volume yields the following:

W_0 = Water added through precipitation to the one-hectare wheat field

W_1 = Water removed by wheat in the one-hectare wheat field

W_2 = Water removed by infiltration or remaining in the one-hectare wheat field

W_3 = Water removed by runoff from the one-hectare wheat field

The amount of water flow onto the one-hectare wheat field is found by

$$W_0 = 0.54 \text{ m/year} \times 10,000 \text{ m}^2 = 5,400 \text{ m}^3/\text{year}$$

A flow balance on the water flowing through the wheat field can be described by

$$W_0 = W_1 + W_2 + W_3$$

Substituting the known percentages of flow distribution into the equation above yields

$$W_1 = (0.61)W_0 = (0.61)\,5{,}400 \text{ m}^3/\text{year}$$

$$W_1 = 3{,}294 \text{ m}^3/\text{year}$$

$$W_3 = (0.22)W_0 = (0.22)\,5{,}400 \text{ m}^3/\text{year}$$

$$W_3 = 1{,}188 \text{ m}^3/\text{year}$$

Using the mass balance equation yields

$$W_0 = (0.61)W_0 + W_2 + (0.22)W_0$$

$$5{,}400 \text{ m}^3/\text{year} = (0.61)\,5{,}400 \text{ m}^3/\text{year} + W_2 + (0.22)\,5{,}400 \text{ m}^3/\text{year}$$

Solving for each unknown term yields

$$W_2 = 918 \text{ m}^3/\text{year}$$

The concentration of the nitrogen in the water leaving the farm field is determined by dividing the mass flow of nitrogen in the runoff by the volumetric flow of runoff water:

$$\text{Concentration} = \frac{\text{Mass flow rate of the substance}}{\text{Volume flow rate of the fluid}}$$

$$\text{Concentration of nitrogen} = \frac{48 \dfrac{\text{kg of nitrogen}}{\text{year}}}{1{,}188 \dfrac{\text{m}^3 \text{ of water}}{\text{year}}} = 0.040 \dfrac{\text{kg of nitrogen}}{\text{m}^3 \text{ of water}}$$

Converting the answer above to the required units,

$$0.040 \frac{\text{kg of nitrogen}}{\text{m}^3 \text{ of water}} \times \frac{1{,}000 \text{ g}}{\text{kg}} \times \frac{1{,}000 \text{ mg}}{\text{g}} \times \frac{\text{m}^3}{1{,}000 \text{ L}} = 40 \frac{\text{mg}}{\text{L}}$$

The concentration of nitrogen in the runoff water leaving this wheat field would be 40 mg/L.

Developing a Process-Based Model

Apply basic engineering principles of problem definition to your system of choice. The assignment goals are to define a system process and related variables and to identify gaps in the knowledge.

5.13 Draw a black box process that illustrates one part of your system of Chesapeake Bay sketched in Problem 5.12.

5.14 Define your process(es) as splitting, mixing, reactions, and so on and show the material flow and/or energy flow streams entering and exiting your process(es).

5.15 Create a list of mathematical variables associated with each flow in the Chesapeake Bay process(es) you defined in the model developed in Problem 5.13.

5.16 Determine the number of variables you will need from the list of variables created in Problem 5.15 to define and properly constrain the process(es) in your system.

5.17 Develop a mathematical model, similar to the system described in Example 5.1, for the Chesapeake Bay process you have modeled in Problem 5.13. Define the relevant mathematical relationships and variables.

- Use peer-reviewed sources or other documented sources to determine values for as many of your controlled variables and uncontrolled variables as possible. (Note: this may over-constrain your system.) Consider starting your search with textbooks, journal articles, government documents, CBF reports, or other resources as appropriate.
- List the variables and reference for the sources of your information; a range of data for a given variable is appropriate.

5.5 Creating a Complex System Model Using Multiple Processes

A system may contain any number of processes or flow junctions, all of which can be treated as black boxes. For example, in the hydrologic cycle, precipitation falls and is added to the watershed into which it falls. At the Earth's surface, some of the rainfall flows out of the watershed, while some of it percolates into the groundwater. If water in the area is used for irrigation, the water is removed from the groundwater reservoir through wells. The irrigation water either percolates into the ground, is incorporated into vegetation, or flows back into the atmosphere through evaporation or transpiration (water released to the atmosphere through plants). Both evaporation and transpiration are commonly combined into one term called "evapotranspiration." This system can be visualized and modeled as a series of black boxes, as shown in Figure 5.10.

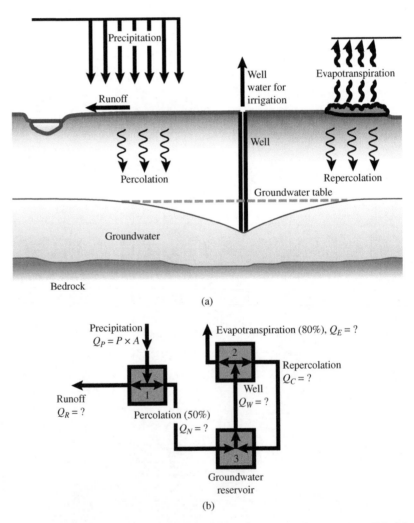

Figure 5.10 (a) Visualization and (b) modeling of the hydrologic cycle as a series of black box processes.

Example 5.2   Sustainable Water Withdrawal for Irrigation

Suppose the rainfall is 40 inches per year, of which 50% percolates into the ground. A farmer irrigates crops using well water. Of the extracted well water, 80% is lost through evapotranspiration; the remainder percolates back into the ground. How much groundwater could a farmer on a 2,000-acre farm extract from the ground per year without depleting the groundwater reservoir volume?

Recognizing this as a material balance problem, first convert the rainfall to a volumetric flow rate; 40 inches per year falling over an area of 2,000 acres yields

$$40 \, \frac{\text{in}}{\text{yr}} \left(\frac{1 \, \text{ft}}{12 \, \text{in}} \right) \times 2,000 \, \text{acres} \left(43,560 \, \frac{\text{ft}^2}{\text{acre}} \right) = 2.90 \times 10^8 \, \frac{\text{ft}^3}{\text{yr}}$$

It is convenient to start the calculations by constructing a balance on the first black box (1), a simple junction with precipitation coming in and runoff and percolation going out. If a fluid is incompressible, such as water, we can conduct a volume flow balance on the fluid material instead of the mass flow balance. It is common in many engineering fields to use the symbol Q to represent the volumetric flow rate of water. The volume rate balance is

Volume rate of water ACCUMULATED	=	Volume rate of water IN	−	Volume rate of water OUT	+	Volume rate of water PRODUCED	−	Volume rate of water CONSUMED

Since the system is assumed to be at steady state, the first term is zero. Likewise, the last two terms are zero, as water, again, is not produced or consumed (e.g., in reactions):

$$0 = \left[2.90 \times 10^8 \, \frac{\text{ft}^3}{\text{yr}} \right] - [Q_R + Q_N] + 0 - 0$$

As stated in the problem, half of the water percolates into the ground; therefore, the other half of the remaining water is runoff:

$$Q_R = 0.5 Q_P = Q_N$$

Plugging this information into the material balance yields

$$0 = 2.90 \times 10^8 - 2 Q_R$$

Solve for Q_R and Q_N:

$$Q_R = 1.45 \times 10^8 \, \frac{\text{ft}^3}{\text{yr}} = Q_N$$

A balance on the second black box (2) yields

$$0 = [Q_W] - [Q_E + Q_C] + 0 - 0$$

As stated in the problem, 80% of the irrigation water is lost by evapotranspiration; therefore, 20% of the irrigation water percolates back into the ground:

$$0 = Q_W - 0.8 Q_W - Q_C$$

$$Q_C = 0.2 (Q_W)$$

Finally, a material balance on the groundwater reservoir (3) can be written if the quantity of groundwater in the reservoir is assumed not to change:

$$0 = [Q_N + Q_C] - [Q_W]$$

From the first material balance,

$$Q_N = 1.45 \times 10^8 \text{ ft}^3/\text{yr}$$

and from the material balance around the second box,

$$Q_C = 0.2Q_W$$

Substituting the information for Q_N and Q_C into the material balance around box 3 and solving yields

$$0 = 1.45 \times 10^8 + 0.2Q_W - Q_W$$

$$Q_W = 1.81 \times 10^8 \text{ ft}^3/\text{yr}$$

This is the maximum safe yield of well water that can be withdrawn from the well by the farmer to irrigate the field.

As a check, consider the entire system illustrated in Figure 5.11 as a black box. The system in question is drawn as Figure 5.11b, and the given information is added to the sketch. Unknown quantities are noted by variables. Precipitation is the only flow into the box. Flow from the box may leave via runoff or evapotranspiration. Representing this black box as Figure 5.11, it is possible to write the material balance in cubic feet per year of water:

$$0 = Q_P - Q_R - Q_E$$

$$0 = (2.90 \times 10^8) - (1.45 \times 10^8) - (1.45 \times 10^8)$$

The balance of the overall system checks the calculations.

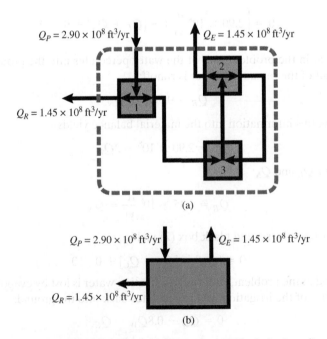

Figure 5.11   Impacts of irrigation from Example 5.2 on the local hydrologic cycle.

When a series of black boxes are combined to model processes within a larger system, it is possible to draw the control volume boundaries around multiple processes within that system, and that larger system of processes can also be modeled as a material balance system. This may be used to check the solution to large complex systems, such as the hydrologic model illustrated in Example 5.2. If the calculations for all the flows are correct within each individual black box, the flows in the larger system should also balance (as they do in Example 5.2).

Developing a Multiple Process-based System Model

5.18 Use your system-level sketch from Problem 5.12 to create a representation of another process connected to the process described in Problem 5.17 of your Chesapeake Bay system.

5.19 Create a list of mathematical variables associated with each flow in the Chesapeake Bay process(es) you defined in Problem 5.18 above.

5.20 Describe each variable listed in Problem 5.19, research each variable, and provide ranges for reported values. Report the values and expected ranges (high, mean, and low values if available). Provide a citation for the source of information for each of the values you find.

5.21 For any unknown variables for which you could not find values reported in the scientific literature, define key words for research studies that might provide information about the unknown variables.

5.22 Describe a system similar to the ones you have chosen to investigate. How might similar systems be studied to help guide any reasonable assumptions you might be able to make in an analysis?

5.23 Describe the boundary conditions of your system and the inherent assumptions required to mathematically define a mathematical model of the system you have described above.

5.24 What related conditions beyond the boundary you defined in Problem 5.18 might impact an analysis of the system? For example, these conditions may be assumptions about the size, shape, environmental conditions, connected flows, or other features of the system you have described. Illustrate these potential conditions and impacts by revising the schematic from Problem 5.18.

Synthesizing Information about Complex Systems

5.25 Synthesize and apply systems thinking, principles, and tools to complex systems. *Sketch the information requested below as a mind map or similar illustration.* The goal is to expand the knowledge about activities in the Chesapeake Bay watershed and illustrate the connection between those activities or processes. You will be asked to peer-review another student's system sketch. The feedback from your peers should be used to improve your individual work on reflection, communication, and refinement of the model you developed in Problems 5.18 through 5.24. Identify how your black box model is linked to five other black boxes being developed by peers in the class. You must speak to more than five different people to find five systems that share a well-defined link.

- Redraw the boundary conditions for your related processes.
- Show all the systems the student interacted with in class in each drawing; some may be linked, while some may not be.
- For each system, define all possible variables that relate to each individual system and identify the names of the class peers related to each system.
- *Illustrate* the variables being analyzed through others' system interactions.
- *Illustrate* the variables that have been ignored by choice in the interacting systems.

5.26 Update the schematic you developed in Problem 5.24 with new information developed from your networking discussion.

- Identify by name the source of information for the *linked black boxes—this creates a new system.*
- Identify known and unknown variables in the system and changes throughout the system of study. Note: This may require additional interactions beyond classroom time, so you might benefit from seeking permission to contact your classmates outside of the normal class period.

5.27 Identify and evaluate trade-offs to make informed decisions about managing resources in the Chesapeake Bay watershed. Is the question you started with in Problem 5.7 the question for which you developed an analytical model? Describe how your thinking about the problem changed throughout the process of developing a model for the system.

5.28 What are the key elements or key words that were most useful in searching for data related to your analytical model development?

5.29 How would policymakers use the information you presented in your Chesapeake Bay watershed model to create policy measures that could be implemented to improve the condition of Chesapeake Bay?

References

Allenby, B. R. (2002). "Earth Systems Engineering and Management." *IEEE Technology in Society* 19, no. 4: 10–24.

Allenby, B. (2007). "Earth Systems Engineering and Management: A Manifesto." *Environmental Science and Technology* 41, no. 23: 7960–7965.

Chesapeake Bay Foundation. (2000). *State of the Bay Report.* Annapolis, MD: Chesapeake Bay Foundation.

Chesapeake Bay Foundation. (2020). *State of the Bay Report.* Annapolis, MD: Chesapeake Bay Foundation.

Gorman, M. E. (2005). "Earth Systems Engineering Management: Human Behavior, Technology and Sustainability." *Resources, Conservation, and Recycling* 44: 201–213.

Gorman, M. E., Allenby, B., Mehalik, M. M., and Hall, K. (2003). "Earth Systems Engineering Management: A Course for Students from Multiple Disciplines." *World Transaction on Engineering Technology Education* 2, no. 1: 29–32.

NASA. (2003). *Earth Science Enterprise Strategy.* NP-2003-01-298-HQ. Washington, DC: NASA Headquarters.

Approaches to Engineering Design

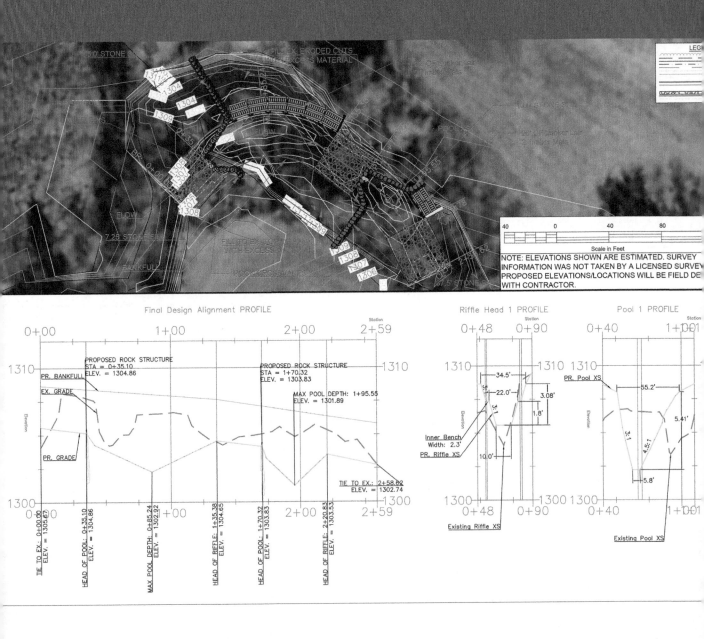

Objectives

Chapter 6 focuses on providing an overview of the engineering design process. Students will develop basic knowledge of the various phases of the engineering design process, including problem definition, conceptual design, embodiment design, and detailed design.

Student Learning Objectives

- Describe various phases of the engineering design process.
- Define a design problem by identifying customer needs.
- Generate multiple design concepts.
- Understand embodiment design.
- Understand detailed design.

6.1   Introduction

Imagine a typical day in your life. From the moment you wake up until the moment you arrive in the classroom, think about all the artifacts—products and services—you use, from the alarm clock on your smartphone to the bus or car you use for transportation. In the very short span of a couple of hours, you may use more than 100 products and services. These can include your air conditioner, toothbrush, shower, coffee maker, toaster, smartphone, laptop, book bag, and so on.

6.1 Of all the products you have used recently, which do you consider your favorite? List all the reasons that make that product your favorite product.

6.2 Of all the products you have used recently, which do you consider your least favorite? List all the reasons that make that product your least favorite product.

Each of these products or services was designed by an engineer. Not only the design, but also the manufacturing and distribution of the products and services were made possible because of design and engineering. Design is an integral part of engineering. Whether it is developing solutions to provide clean drinking water in rural areas or building self-driving cars, design is central to devising these solutions.

In the remainder of this chapter, you will learn about the engineering design process and various phases associated with the engineering design process.

6.2   Engineering Design Process Overview

The engineering design process is a series of systematic steps that allows an engineer to explore a given design problem, identify pain points, and systematically work toward finding an appropriate solution. There are many different models of the design process, including the engineering design process, IDEO's human-centered design process, and Stanford d-school's design thinking process.

6.3 Look up and draw the schematic of the various phases of the engineering design process (www.teachengineering.org/populartopics/designprocess).

6.4 Look up and draw the schematic of the various phases of the human-centered design process as prescribed by IDEO (www.ideo.org/tools).

6.5 Look up and draw the schematic of the various phases of Stanford d-school's design thinking process (empathizeit.com/design-thinking-models-stanford-d-school).

6.6 Write down the similarities and differences you observed after comparing (1) the engineering design process, (2) IDEO's human-centered design process, and (3) Stanford d-school's design thinking process.

Even though these design processes have some similarities and differences, each process ultimately allows an engineer to explore a given design problem, identify pain points, and develop an appropriate solution for the problem.

For our purpose, we will consider four phases of the engineering design process. While the engineering design process can be used for product, process, or service design, we will focus primarily on product design. Various tools and methods are used during each phase of the design process to guide the engineer through each of these phases. These are briefly described below:

1. **Problem definition phase:** This is the first phase of the engineering design process and is focused on exploring and defining the design problem at hand. Often, this problem definition phase starts with an open-ended design problem, focuses on identifying the pain points of the user, and ends with a clear definition of requirements that must be fulfilled to solve the given problem. Examples of tools and methods used during the problem definition phase include preliminary research, benchmarking, reverse engineering, interviews to gather user needs, personal development, and the elicitation of requirements.

2. **Conceptual design phase:** After defining the design problem and gathering requirements, the conceptual design phase is the next phase of the engineering design process. It focuses on exploring the design space, considering multiple solutions for the problem, and prioritizing the best concepts. Typically, the conceptual design phase starts with a relatively well-defined design problem and requirements and ends with a list of top concepts. Examples of tools and methods used during the conceptual design phase include concept generation tools such as brainstorming, C-sketch, morphological matrix, and the gallery method, as well as concept evaluation and selection tools such as decision matrix, the Pugh method, and the analytical hierarchy process.

3. **Embodiment design phase:** The third phase in the engineering design process is the embodiment design phase. In this phase, the focus is on embodying the concepts selected at the end of the conceptual design phase. This is done by creating multiple iterations of prototypes and testing and refining the prototypes. Additionally, various analyses are also done to check the feasibility and robustness of various design solutions. Examples of tools and methods used during the embodiment design phase include physical prototyping, digital prototyping using computer-aided design software, design for manufacturing (DFM), design for assembly (DFA), and design of experiments (DOE).

4. **Detailed design phase:** The detailed design phase is the fourth phase in the engineering design phase. In this phase, the design is brought to the state of completion whereby it is ready for production. Any final kinks in the design are fixed, and various details pertaining to the dimensions, materials, and manufacturing process are added to the design. Additionally, final drawings and cost estimates are also completed. Typically, a final design review is conducted before sending the design out for production. Examples of tools and methods used during the detailed design phase include engineering drawings, final design specifications, and cost estimation.

It is important to note that the design process rarely progresses linearly; rather, it is iterative in nature. This means that an engineer might have to go back to the problem definition phase after performing concept generation to further understand the design problem. At each iteration, new knowledge is acquired, and further improvements are made going forward.

6.3   Problem Definition Phase

The engineering design process starts with the problem definition phase. Engineering problems often arise due to an unfulfilled need of a user or due to current unsatisfactory performance of an existing solution.

Most design problems are vaguely defined and open-ended. During the problem definition phase, the goal is to gather as much information as possible so that the open-ended vague problem can be defined more clearly. Following are some ways to gather preliminary information about the design problem:

1. Literature review: reviewing articles, book chapters, magazines, handbooks, and so on to further understand the design problem. While conducting a literature review, it is important to review only material that is from credible sources.

2. Benchmarking: reviewing existing solutions for the design problem and understanding what works well and what does not work well. For example, if you are tasked with designing a new water bottle, you may want to benchmark some of the top-selling water bottles in the market to understand what needs are still unfulfilled.

3. Reverse engineering: typically done to redesign a current solution and make improvements. For reverse engineering a product, you dismantle the existing product part by part, understand the functionality of each part, and look for areas of improvement.

In addition to preliminary and background information, it is also important to understand the needs and wants of the user. *A user is any individual who will end up using the final artifact that is designed.* Often, this may or may not be the individual who pays for the artifact. For example, if you are designing a new toy for a toddler, even though the toddler is the ultimate user of the product, the parents are the ones who will pay for the toy and are equally important users for the product. In order to understand and define a design problem, it is therefore important to understand the needs (requirements) of all the users. There are many techniques for gathering the needs of the user; here we will focus on interview as a way to collect the needs of the user.

Interview as a Method to Gather User Needs

Interviewing the user allows us to gather the needs of the user in terms of their pain points, likes, and dislikes. In order to gather useful information from the user that will ultimately help define the problem, it is important to ask good open-ended questions. For example, if you are designing new features for a bicycle, instead of asking the user, "Do you like to ride a bicycle?", which may result in a simple yes-or-no answer without giving much further information, the engineer should instead ask a more open-ended question, such as, "What aspects about riding a bicycle do you like or dislike?" This question will result in more useful information that will allow the engineer to understand the pain points of the user.

Before conducting the interview to gather user needs, the engineer must come up with a list of questions to ask the user. Select a time and location convenient for the user. Avoid using jargon while interviewing the user. Remember, the user may not be a technical expert. The interview should not typically last for more than 30 to 45 minutes. You may need to interview multiple users to capture all the needs. Take notes during the interview to capture the needs of the user. If you are planning to use audio or video to record the interview, it is important to ask for the permission of the user ahead of time. It is okay to ask follow-up questions if required. Finally, thank the user for their time at the end of the interview.

Defining Requirements

At the end of the user interviews, you should have collected notes of user responses. These notes will then allow you to understand the pain points, likes, and dislikes of each user. However, these notes are often written as raw user statements as collected during the interview and must be rewritten as design requirements.

Design requirements are needs of the user expressed more formally and, when possible, with precise measurable details. Some examples of converting the user raw statements for a shopping cart redesign to design requirements are shown below:

User Raw Statements	Design Requirement
1. "It would be awesome if my shopping cart would adjust to my height, as I am a relatively short person."	The shopping cart must have variable height options between four and six feet.
2. "I would really like my shopping cart to provide some kind of entertainment for my toddler while I shop."	The shopping cart should provide some kind of entertainment for toddlers.
3. "Often times, the wheels of the shopping cart are very squeaky and make for a very annoying shopping experience."	The shopping cart must have a smooth ride throughout its life span.

Exploring the Design Problem

Design Prompt   Do you carry a book bag to college? Do you absolutely love it or hate it, or are you somewhere in between? Does it have too many pockets or not enough pockets? Does it accommodate your needs of carrying a laptop, iPad, snacks, keys, smart devices, and so on, or does it barely have space for books? Is it smart and sleek or bulky and ugly?

You have been hired by FutureBookBag Inc. to design a book bag for the future. Most book bags currently in the market are bulky and aesthetically unappealing. Some have too many pockets to keep track of, while others are difficult to carry and cause back pain. Further, these book bags were not designed keeping in mind the needs of 21st-century college student. You are to design a book bag catering to the needs of 21st-century college students and costing no more than $50.

6.7 Write down the initial problem statement based on your understanding of the design prompt. The problem statement must include the current undesirable state and desired goal state.

6.8 Identify potential users. This could be an individual who will either pay or use your final product.

6.9 List five to seven questions that you would want to ask the user during the interview. Remember to frame open-ended questions that will allow you to gather the pain points, likes, and dislikes of the user.

6.10 Jot down the notes from the interview. If you conduct multiple interviews, jot down the notes for each interview separately. Look for similarities and differences and capture the likes and dislikes for all users you interviewed.

6.11 Based on the notes gathered during the interview(s), write down an updated problem statement to reflect the needs of the user.

6.12 Based on the notes gathered during the interview(s), list at least 10 unique requirements.

6.4 Conceptual Design Phase

After the problem definition phase, the next phase in the engineering design process is the conceptual design phase. The conceptual design phase involves concept generation, concept evaluation, and concept selection.

Concept generation involves generating many different concepts for the given design problem. In order to allow a thorough exploration of the design space, many different concept generation techniques are developed. It is recommended to use multiple methods while generating concepts so that different aspects of the design space can be explored. Here we will focus on brainstorming as a method for concept generation.

Brainstorming as a Concept Generation Method

Brainstorming is a concept generation method that allows us to generate a large quantity of concepts in a short amount of time. There are specific rules to brainstorming, but they may be modified to suit the needs and resources available. The rules of brainstorming are as follows:

1. Brainstorming can be performed in teams of a minimum of two individuals to a maximum of 15 individuals. More than 15 individuals can make the brainstorming process too chaotic.

2. One individual may be designated as the recorder of ideas and may not participate in the actual brainstorming.

3. The brainstorming session starts by writing down the design problem and making sure that everyone participating in the session has the same understanding of the problem.

4. After the problem is written, the participating members can freely share their ideas. There should be no judgment or discussion on the ideas being shared. The recorder must make sure to record all the ideas.

5. The participants can take turns sharing their ideas or can freely share ideas as they come. It is important to not interrupt another person or judge any ideas at this stage.

6. One brainstorming session should not last for more than 60 minutes. If another session is required, it may be scheduled at another time.

7. At the end of the brainstorming session, all the ideas must be recorded. These ideas may be represented by sketches or written descriptions.

Concept evaluation involves assessment of the design concepts generated during concept generation against the requirements developed in the problem definition phase. For concept evaluation, the different concepts are ranked against the requirements based on the extent to which they address or meet the requirements.

Concept selection involves selecting the top concepts from the pool of concepts evaluated during concept evaluation. These concepts will then be further developed through prototyping.

Exploring the Conceptual Design Phase

Now that you have a better understanding of the design problem, we will focus on exploring the conceptual design phase.

6.13 Using brainstorming as a method to generate concepts, generate and document (as either a sketch or a written description) at least 20 unique concepts for the futuristic book bag.

6.14 From the concepts generated in Problem 6.13, select the top eight and sketch them out in more detail below. You may also consider including a brief description (two or three sentences) for each concept.

Sketch 1: Description:	Sketch 2: Description:
Sketch 3: Description:	Sketch 4: Description:

Sketch 5:

Description:

Sketch 6:

Description:

Sketch 7:

Description:

Sketch 8:

Description:

6.15 From the eight concept sketches developed in Problem 6.14, gather some feedback from the users and narrow the list down to only four concepts. Perform concept evaluations and selection by comparing each of the four remaining concepts against the 10 requirements you developed in Problem 6.10. Use a ranking scale of 1–5, where

- 1 = does not meet the requirements,
- 2 = meets a few requirements,
- 3 = meets some requirements,
- 4 = meets most requirements, and
- 5 = fully meets the requirements.

At the end of this activity, select the top two concepts for the next design phase.

Design Requirement	Concept 1	Concept 2	Concept 3	Concept 4
1				
2				
3				
4				
5				
6				
7				
8				
9				
10				
Total Score = Add the ranking for each concept.				

6.16 Clearly sketch and describe the top two concepts as identified in Problem 6.15 based on the scores.

Top Concept 1 Sketch and Description	Top Concept 2 Sketch and Description

6.5 Embodiment Design Phase

The third phase in the engineering design process is the embodiment design phase. In this phase, the focus is on embodying the concepts selected at the end of the conceptual design phase. This is done by creating multiple iterations of prototypes, testing, and refining the prototypes. Typically, the top two or three designs are prototyped. While many different activities are performed during the embodiment design phase, as mentioned above, we will focus on prototyping to explore the embodiment design phase.

Prototyping

Prototypes are the "manifestation of an idea into a format that communicates the idea to others or is tested with users, with the intention to improve that idea over time" (McElroy 2017).

Based on this definition, one of the goals of creating a prototype is to communicate the idea to the user in a tangible way. A user can interact with a physical prototype and provide feedback on the look, feel, and functionality of the solution. This feedback from the user can then be used to further improve and iterate on the solution.

Prototypes are often simple representations of the design and therefore can initially be made with readily available materials such as cardboard, foam board, popsicle sticks, duct tape, and so on. More refined prototypes may require the use of maker space, machine shops, or fabrication studios and the use of skills such as woodworking, three-dimensional printing, laser cutting, machining, and so on. The decision about the type of prototype to create will depend on the time and budget available, the skill set within the team, and the purpose of the prototype.

Exploring the Embodiment Design Phase

Now that you have selected your top two concepts, let us explore the embodiment design phase.

6.17 For each of your top two concepts, list the materials you plan to use for making initial prototypes.

6.18 For each of your top two concepts, list the processes or manufacturing techniques needed to make an initial prototype. Examples include three-dimensional printing, sewing, woodworking, and so on.

6.19 Build a prototype of each of your top two concepts in a way that a potential user can interact with and provide feedback to. In the space below, write down the feedback you got from the users for each of the two prototypes.

6.20 What design decisions might you make based on the feedback you got for your prototypes? Will you change anything in your design? If yes, what and why? If no, why not?

6.6 Detailed Design Phase

The detailed design phase is the fourth and final phase in the engineering design process. In this phase, the design is brought to the state of completion whereby it is ready for production. The final kinks in the design are fixed, and various details pertaining to the dimensions, materials, and manufacturing process are added to the design. Additionally, final drawings and cost estimates are also completed. Typically, a final design review is conducted before sending the design out for production.

During the detailed design phase, final engineering drawings are completed using computer-assisted design (CAD) software. These drawings typically have all dimensions and final design specifications. Additionally, the final production details as well as cost estimations are complete at this stage. The final design is then sent for production.

Exploring the Detailed Design Phase

Now that you are in the last phase of the engineering design process, let us explore the detailed design phase by thinking about the final details for the book bag design.

6.21 Draw a detailed hand-drawn sketch of your final book bag design with all major dimensions labeled.

6.22 Research and identify the materials you would like to use for the final book bag design. Justify your decision of selecting these particular materials.

6.23 Research and identify some manufacturing processes you might need to use to make the final book bag design. List them below.

6.24 Research and identify off-the-shelf components that are available for purchase and list the cost of each component. If off-the-shelf components are not available, research and identify raw materials and the cost of each material required to produce one backpack. Add the cost of off-the-shelf components and raw materials to estimate the material cost of one backpack. How does this compare to the $50 target for your design? Comment on the material costs and potential profit margins; do not forget to consider labor costs and seller expenses in your comments. Should you promote this first design to FutureBookBag Inc., or is it back to the drawing board?

References

Dieter, G. E., and Schmidt, L. C. (2021). *Engineering Design*. New York: McGraw-Hill.

McElroy, K. (2017). *Prototyping for Designers: Developing the Best Digital and Physical Products*. Sebastopol, CA: O'Reilly Media.

Objectives

Chapter 7 focuses on understanding energy sources, energy use, and implications of energy flow in the engineering processes. Students will develop basic knowledge and tools to identify and differentiate between fossil fuels and alternative fuel sources. They will also identify the implications of sources of energy in energy portfolios available to the designer, develop an understanding of the implications of energy choice on climate systems, and investigate energy options.

Student Learning Objectives

- Identify the role engineers play in energy development.
- Calculate energy efficiency.
- Describe how engineers consider embodied energy in design.
- Demonstrate how greenhouse gas reductions can be achieved through energy choices.
- Describe how energy portfolio choices might be considered to develop future energy policy.

7.1 Introduction

Energy development and supply is one of the primary engineering tasks associated with the sustainable development goals. Sustainable Development Goal (SDG) number 7 is to ensure access to affordable, reliable, sustainable, and modern energy. Development of reliable and efficient energy production relates to SDG number 3 (to provide good health and well-being) and number 8 (to ensure decent work and economic growth by providing energy for homes, industry, schools, and health care). Energy for lighting and internet access is needed for studying and equal access to information. Our modern infrastructure also requires energy for modern water and sanitation processes, traffic lights for transportation systems, and energy distribution systems. Responsible consumption and production, SDG number 12, might include integrating solar energy into roofing technologies and other building technologies. The Earth's climate action (SDG number 13) is directly impacted by energy use and energy consumption of fossil fuels and could be mitigated by developing cleaner energy sources. As low- and moderate-income countries increase their energy demand, more efficient energy portfolios are essential for providing infrastructure that provides access to energy and other modern-day conveniences.

7.2   Energy and Mechanical Engineering

Energy development and energy efficiency improvements have long been the province of engineers, especially mechanical engineers. Mechanical engineers are often stereotyped as automobile and vehicle engineers; however, mechanical engineering is a much, much broader field. It is true that automobiles and the transportation sector consume over a third of the energy produced in the United States, most of which is from petroleum-based fossil fuels. The transportation sector is responsible for almost a third of greenhouse gas emissions in the United States; therefore, decreasing greenhouse gas emissions from automobiles is an important engineering initiative. It is much less commonly known that mechanical engineers design the power plants and energy distribution systems used in the United States. In 2021, global energy-related carbon dioxide emissions exceeded 36.3 pentagrams (36.3×10^{15} g), much of which were from fossil fuels. This amounts to 4.6 metric tons of carbon dioxide emitted for every living person! Developing more efficient energy systems is necessary to limit the negative impacts of the warming climate. Mechanical engineers have a leading role in developing renewable energy technologies, such as solar panels and wind turbines. Mechanical engineers must design systems to meet the energy demands of the future by utilizing new technologies.

7.1 Describe how the U.S. Bureau of Labor and Statistics defines a mechanical engineer, the entry-level education required, and the median salary for mechanical engineers by reviewing the Field of Degree highlights for students and job seekers in the *Occupational Outlook Handbook*.

7.2 Describe how the U.S. Bureau of Labor and Statistics defines a petroleum engineer, the entry-level education required, and the median salary for a petroleum engineer by reviewing the Field of Degree highlights for students and job seekers in the *Occupational Outlook Handbook*.

7.3 Find job listings for a mechanical engineering position, a petroleum engineering position, or a nuclear engineering position. Describe the similarities and difference between the education required, experience required, and salary range (if available) for two different positions. What other educational degrees or requirements are acceptable for the job listing you have found? You can find job listings through employment websites, USAJobs.gov, ASME.org, LinkedIn, and other professional networking websites.

7.3 Energy Consumption

Fossil fuels, including coal, petroleum, and natural gas, are still the primary source of energy in the United States, as shown in Figure 7.1. Non-carbon-based energy sources make up an increasing proportion of the energy supply in the United States, as illustrated in Figure 7.2. These energy sources are used to create the electric energy that powers the transportation, industrial, residential, and commercial sectors needed by our modern-day civilization. Energy and electrical production come from a variety of sources typically delivered through or connected with regional energy providers. Energy companies develop energy portfolios that form a variety of energy sources used to meet the demand for power in homes and industry. The energy portfolio may be developed based on the costs and availability of various sources, legal policies and mandates that set requirements for a certain proportion of energy sources to be present in the portfolio, negotiated long-term contracts with energy providers, and other incentives.

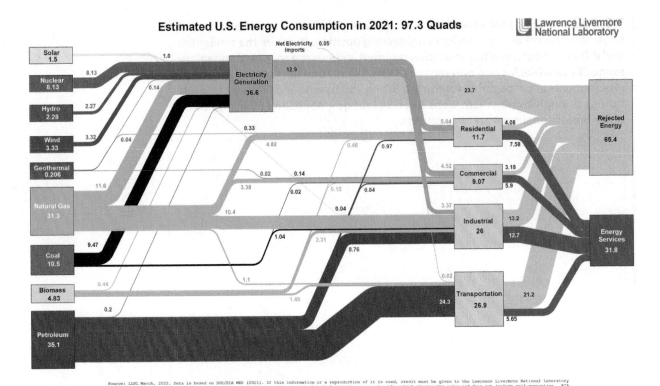

Figure 7.1   Estimated U.S. energy consumption by sources and sectors in 2021.
Source: flowcharts.llnl.gov.

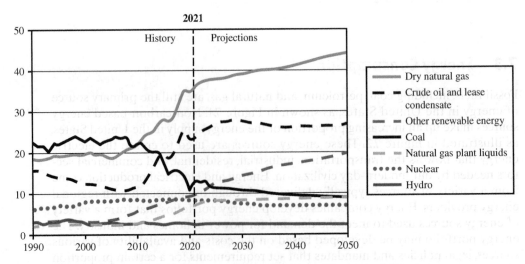

Figure 7.2   Energy production in the United States by source in quadrillion Btus.
Source: EIA Energy Outlook 2022, www.eia.gov/outlooks/aeo/pdf/AEO2022_ChartLibrary_full.pdf.

Examining Your Energy Portfolio

7.4 Estimate how much energy you use each day.

a) From your mileage traveled by car, motorcycle, or public transportation and your apartment or home energy bill (or a sample energy bill provided by the instructor), determine the amount of kilowatt-hours of energy used on average each month.

b) Show the list of transportation, industrial use, residential use, and electric power consumption in a bar or pie chart.

c) Determine from your regional energy provider the source of your energy: coal, natural gas, petroleum, renewable energy, nuclear, and so on. Visit spotforcleanenergy.org for your regional energy portfolio. Show the portfolio of energy used by source in a bar or pie chart.

d) How closely did your graph estimates match the sources of energy in your state?

e) From the Gap Analysis on the State Policy Opportunity Tracker (SPOT) for clean energy, list three to five energy market opportunities for which your state is best prepared for its energy future and list three to five energy market opportunities for which your state is least prepared for its energy future. Briefly summarize why your state is well prepared for some energy opportunities and much less prepared for other energy opportunities.

7.4 Energy Units

The units of energy used in the Système International d'Unités, or the *SI* system, is the Joule, and it is equivalent to the units used to define the work done on a system where

$$1 \text{ Joule (J)} = 1 \text{ Newton-meter} = 1 \frac{\text{kg} \cdot \text{m}^2}{\text{s}^2}$$

The SI unit for thermal energy is the calorie, which is historically based on the thermal energy required to increase the temperature of 1 g of water by 1°C (from 14.5°C at a pressure of 1 atm). Power is defined as the work divided by time, or the rate at which energy is converted to power. The unit of measure is the watt, where

$$1 \text{ watt (W)} = 1 \frac{\text{J}}{\text{s}}$$

Many systems in the United Kingdom and the United States still use the *American engineering system* units for energy. The British thermal unit (or Btu) has a similar basis to the calorie; it is defined as the amount of energy required to raise one pound of water by 1°F. The conversion factors for common energy units are shown in Table 7.1.

Table 7.1 Energy conversion units

Commonly Used Quantities of Energy and Energy Content	
1 calorie (cal)	4.184 J
1 British thermal unit (Btu)	1,055 Joules (J)
1 megajoule (MJ)	948 Btu
1 kilowatt-hour (kWh)	3,412 Btu
1 horsepower (hp)	746 watts
1 barrel of crude oil (bbl)	42 U.S. gallons = 5,100,000 Btu
1 ton of oil equivalent (toe)	41.868 GJ
1 therm	100 ft³ natural gas (ccf) = 103,100 Btu
1 gallon of gasoline	125,000 Btu
1 gallon of no. 2 fuel oil	138,690 Btu
1 gallon of LP gas	95,000 Btu
1 ton of coal	25,000,000 Btu
1 ft³ natural gas	1,031 Btu
1 MMBtu	1 million Btu
1 quad	10^5 Btu
(short) ton	2,000 lb
long ton	2,240 lb

The sources of energy shown in Figures 7.1 and 7.2 have different amounts of energy available per unit mass of the source, called the energy densities. For example, wood has an energy density that is typically in the range of 13 to 19 megajoules per kilogram compared to hard (anthracite) coal, which has an energy density of about 25.8 megajoules per kilogram, as shown in Table 7.2. Thus, a smaller mass and volume of anthracite coal has a greater potential for energy than wood. This explains why the United States has derived much of its electricity production from coal rather than from wood (or other biofuels) since a much greater amount of electricity can be produced from a train carload of coal than from a train carload of wood. Fuel oil no. 2, which is derived from petroleum distillation, has historically been used for home heating oil, as the fuel oil is almost twice as energy efficient as coal and almost four times more energy efficient than wood.

7.5 Direct and Embodied Energy

We can analyze and account for energy use as direct or as embodied energy, and each approach gives us important insights into energy consumption patterns. ***Direct energy*** represents activities that involve the conversion of an energy fuel or resource into usable energy. When we use gasoline to power a car, that is direct energy use. Figure 7.3 shows the U.S. Energy Information Administration data on direct energy consumption from different sources used in U.S. agriculture in 2014. Other examples of direct energy use include converting sunlight with photovoltaic cells

Table 7.2 Energy density of common fuels

Fuel	Energy Density (MJ/kg)
Peat	10.4
Wood, oak	13.3–19.3
Wood, pine	14.9–22.3
Coal, anthracite	25.8
Charcoal	26.3
Coal, bituminous	28.5
Fuel oil no. 6 (bunker C)	42.5
Fuel oil no. 2 (home heating oil)	45.5
Gasoline (84 octane)	48.1
Kerosene	45.6
Natural gas (density = 0.756 kg/m³)	53.0

Source: Based on Davis, M.L., and Masten, S.J., *Principles of Environmental Engineering and Science* (New York: McGraw-Hill, 2004).

into electricity and burning natural gas to produce steam for an industrial process. Basically, if we are burning a fuel or generating electricity, we are engaged in direct energy use. Figure 7.1 provides an understanding of direct energy use and the flow of fuels by end-use sectors in the United States in 2021.

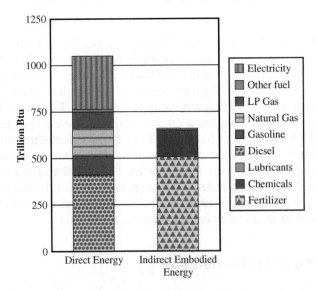

Figure 7.3 Direct and indirect energy sources used in the U.S. agriculture industry in 2014. Energy makes up a significant part of operating expenditures for most crops, especially when considering indirect energy expenditures on fertilizer, because the production of fertilizer is extremely energy intensive and requires large amounts of natural gas.

Source: Claudia Hitaj and Shellye Suttles, *Trends in U.S. Agriculture's Consumption and Production of Energy: Renewable Power, Shale Energy, and Cellulosic Biomass*, EIB-159 (Washington, DC: U.S. Department of Agriculture, Economic Research Service, August 2016).

Changes to the U.S. Energy Portfolio since 1800

7.5 Describe the technological and engineering changes that resulted in changes to the sources of energy used in the United States during each century listed below. The resources at the U.S. Energy History Visualization project (us-sankey.rcc .uchicago.edu) may be helpful in your analysis. The interactive visualization shows 200 years of evolving energy use in the United States as an animated Sankey diagram. Line widths represent per capita energy flows each year from primary energy sources (left) to final uses (right). The project is an effort of the University of Chicago's Center for Robust Decision-Making on Climate and Energy Policy (RDCEP), and its research is documented in Suits, Matteson, and Moyer (2020).

a) What was the primary energy source used in the United States in 1800? What energy sector was the largest consumer of energy in 1800? Why was this the energy source of choice? Describe the energy content of the primary source as high, medium, or low.

b) What was the primary energy source used in the United States in 1900? What energy sectors expanded from 1800 to 1900? What changes in technology created a demand to change the source of energy? Why was this the energy source of choice? Describe the energy content of the primary source as high, medium, or low. Describe how the change in overall energy use and sources may have impacted pollution (including greenhouse gases) compared to 1800.

c) What was the primary energy source used in the United States in 2000? What energy sectors expanded from 1900 to 2000? What changes in technology created a demand to change the source of energy? Describe the energy content of the primary source as high, medium, or low. Describe how the change in overall energy use and sources may have impacted pollution (including greenhouse gases) compared to 1900.

d) What do you believe will be the primary energy source used in the United States in 2100? What changes in technology are most likely to result in changes to the source(s) of energy? Describe how the change in overall energy use and sources may impact pollution (including greenhouse gases) compared to the year 2000.

Embodied energy, also referred to as indirect energy, is a different kind of energy accounting. It represents the total amount of energy required to make and transport a finished product or material. Embodied energy includes the energy needed to extract and process natural resources into usable material feedstocks, to manufacture the product, and to ship items from place to place as the product moves through its life cycle. Embodied energy represents the sum of all the direct energy that was used to create and transport a specific product. Figure 7.3 shows the importance of embodied energy in agriculture. Fertilizer manufacturing utilizes large amounts of natural gas that is characterized along with other energy-intensive processes as embodied energy in agricultural production.

Laws of physics govern the potential contributions of embodied energy to sustainability. Newton's first law of thermodynamics postulates that all energy is conserved in a closed system. However, because of entropy, the second law tells us that not all energy can be recovered because some processes are irreversible. For example, the energy initially contained in a fossil fuel cannot be completely recovered and used again after it has been burned. The concept of embodied energy also enables us to consider the overall productivity of life cycle energy for an energy source.

Embodied energy is accounted for using life cycle analysis. This sort of analysis allows us to think more holistically about energy sustainability. By considering where and how products consume energy throughout their life cycle, we can target the most energy-intensive stages for decarbonization efforts. Embodied energy also allows us to think about the role of recycling in energy conservation and our ability to recover the energy used for our manufactured products. Recycling avoids the energy required to extract and process a natural resource for first use. Embodied energy is recovered and used when, for example, waste-to-energy facilities burn trash to make process steam or electricity.

7.6   The Carbon Footprint

Energy sources and uses in the United States are shown in Figure 7.1. Information about energy use and fossil fuel consumption can be used to calculate the carbon emissions associated with the conversion of fossil fuels to energy. In general, the combustion of fossil fuels can be described by

$$\text{Fuel} + \text{Oxygen} \rightarrow \text{Energy} + \text{Carbon dioxide} + \text{Water}$$

Depending on the chemical composition of the fossil fuel, differing amounts of energy and carbon dioxide are produced. Carbon dioxide and other gases that increase the energy absorbed by the atmosphere, called greenhouse gases (GHGs), are emitted from anthropogenic and natural sources. Approximately 73.2% of greenhouse gas emissions in 2016 were due to energy uses, such as electricity production, heating, and cooling. Agriculture and land use emissions were the second-highest source of greenhouse gas emissions in 2016 and accounted for 18.4% of total emissions. Carbon dioxide is removed by plant growth, particularly in densely forested areas; the deforestation due to changes in land use for agriculture or the built environment accounted for a 2.2% increase in greenhouse gas emissions in 2016. Energy use in built infrastructure (17.5%) and transportation (16.2%) produce about one-third of all greenhouse gases. Iron and steel production (7.2%), chemical and petrochemical production (3.6%), and cement production (3%) are the largest industrial contributors to GHGs in the atmosphere. The fossil fuels do not contribute equally to GHGs because of differences in their chemistry and because the combustion equipment that burns them have different efficiencies. These CO_2 emission factors are not the same throughout the world; they are calculated by country because of differences in each nation's fuel mix and types of combustion technologies. Figure 7.4 illustrates the flow of GHGs in the United States from their sources, through their use, and to emitted gases at the end of use.

Carbon footprints are the primary accounting method for tracking and analyzing GHG emissions from sources or activities and enable us to evaluate decarbonization efforts. The concept of embodied energy allows us to look at the

U.S. GHG Emissions Flow Chart

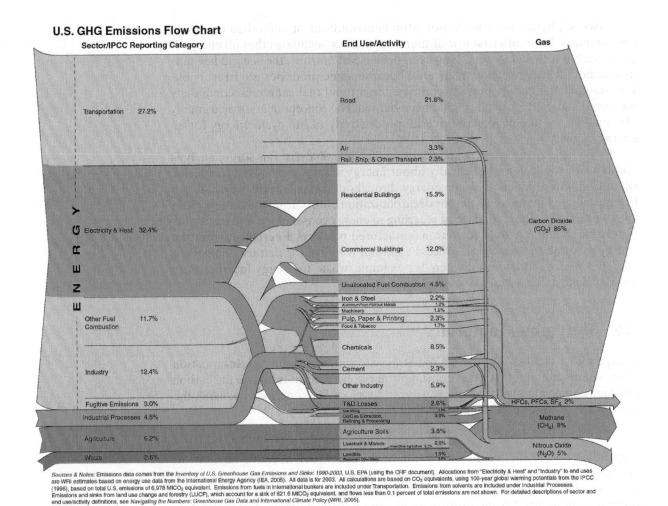

Sources & Notes: Emissions data comes from the *Inventory of U.S. Greenhouse Gas Emissions and Sinks: 1990-2003*, U.S. EPA (using the CRF document). Allocations from "Electricity & Heat" and "Industry" to end uses are WRI estimates based on energy use data from the International Energy Agency (IEA, 2005). All data is for 2003. All calculations are based on CO_2 equivalents, using 100-year global warming potentials from the IPCC (1996), based on total U.S. emissions of 6,978 $MtCO_2$ equivalent. Emissions from fuels in international bunkers are included under Transportation. Emissions from solvents are included under Industrial Processes. Emissions and sinks from land use change and forestry (LUCF), which account for a sink of 821.6 $MtCO_2$ equivalent, and flows less than 0.1 percent of total emissions are not shown. For detailed descriptions of sector and end use/activity definitions, see *Navigating the Numbers: Greenhouse Gas Data and International Climate Policy* (WRI, 2005).

Figure 7.4   This flowchart shows the sources and activities across the U.S. economy that produce greenhouse gas emissions. Energy use is responsible for the majority of greenhouse gases. Most activities produce greenhouse gases both directly, through on-site and transport use of fossil fuels, and indirectly, from heat and electricity that comes "from the grid."
Source: www.wri.org/data/us-greenhouse-gas-emissions-flow-chart.

overall impact of specific products on energy use and GHGs. As previously noted, embodied energy is the *sum of all the energy* needed to produce a product, as if that energy was incorporated into the product itself. The carbon footprint is the *sum of all the greenhouse gases* emitted by the full life cycle of the product.

Carbon footprint techniques are used to calculate GHG emissions for different energy activities as well as the emissions that are *avoided* by implementing carbon mitigation efforts. Institutions and policymakers can also use the carbon footprint to track changes in their GHG emissions by calculating their carbon footprint for a baseline year. The potential climate benefits of investing in resource conservation and green technologies can be determined. For example, companies that adopt a renewable energy portfolio can calculate the measurable benefits of investing in forest protection, planting trees to increase carbon sequestration, or investing in renewable energy.

Table 7.3   U.S. Environmental Protection Agency (EPA) carbon dioxide emission factors for house-hold resource consumption

Source	Source Units	CO_2 Conversion Factor	Conversion Factor Units
Electricity	kWh	5.6178×10^{-4}	Tonnes CO_2/kWh
Coal	Railcar	232.74	Tonnes CO_2/90.89 tonnes railcar
Natural gas	ft³	5.42×10^{-5}	Tonnes CO_2/ft³
Fuel oil	Barrel	0.43	Tonnes CO_2/barrel
Gasoline	Barrel	0.1778	Tonnes CO_2/barrel
Kerosene	Barrel	0.42631	Tonnes CO_2/barrel
Propane cylinders used for home	20-lb cylinder	0.0265	Tonnes CO_2/cylinder
Uptake by trees	Urban tree planted	0.039	Tonnes CO_2/tree
U.S. forests storing carbon for one year	One acre of average U.S. forest	1.22	Tonnes CO_2/acre-year

Source: Based on U.S. EPA, *Clean Energy Calculations and References*, www.epa.gov/cleanenergy/energy-resources/refs.html, accessed October 12, 2022.

Calculating the Carbon Footprint of an American Household

7.6 The U.S. EPA (2013) reported that, on average, an American home consumed the following energy resources. The U.S. EPA uses a simplified method of estimating an individual's greenhouse gas emissions in their Household Carbon Footprint Calculator (www3.epa.gov/carbon-footprint-calculator). The carbon dioxide equivalent emission rates are estimated from the conversion factors in Table 7.3. What is the simplified carbon footprint for average home energy use in the United States? To answer this question, the individual GHG emissions for each category are calculated and then summed to estimate the total carbon footprint for an average U.S. household.

Energy Source	Amount Consumed
Delivered electricity	11,319 kWh
Natural gas	66,000 ft³
Gasoline	464 gallons
Fuel oil	551 gallons
Kerosene	108 gallons
Auto transportation (gasoline)	Two vehicles at 11,398 miles each
Auto fuel economy	21.6 mpg

a) Calculate CO_2 emissions in metric tonnes per year for electricity from the home by multiplying the U.S. EPA carbon dioxide emission factor for household resource consumption by the amount of electricity consumed.

b) Calculate CO_2 emissions in metric tonnes per year for natural gas use from the home.

c) Calculate CO_2 emissions in metric tonnes per year for home gasoline use. There are 20 gallons of gasoline in a barrel.

d) Calculate CO_2 emissions in metric tonnes per year for fuel oil use from the home. There are 42 gallons of oil in a barrel.

e) Calculate CO_2 emissions in metric tonnes per year for kerosene gas use in the home.

f) Calculate CO_2 emissions in metric tonnes per year for transportation via automobile.

g) Calculate the *total* CO_2 emissions in metric tonnes per year for a typical home in the United States.

7.7 Decarbonization of Energy Supply

Fossil fuel use for energy supply has led to the technological changes that allow modern societies to produce food and goods, distribute food and goods through modern transportation, heat and cool homes, and generally provide many of the services on which modern society is based. However, these same fossil fuels are primarily responsible for rapid climate change that threatens food supply, water supply, and many of the benefits of our modern society. Climate models predict food and water shortages for millions of people, severe weather-related disasters, and mass extinctions of species if the average surface temperature of the planet warms beyond 1.5°C or 2°C. Limiting global warming to either 1.5°C or 2°C relies on major changes to the energy system. The task is daunting, and a swift transition to net-zero global carbon emissions requires that we work largely within the existing system and in anticipation of growing energy needs in many parts of the world.

These considerations include the technologies and infrastructures designed around fossil fuels, the impacts of economic development on energy consumption, the different types of energy services and derived demand created by the end-use sectors, the special needs of low-income countries, and the availability of energy alternatives to fossil fuels. Blueprints for deep decarbonization are detailed by the International Energy Agency in *Net Zero by 2050: A Roadmap for the Global Energy Sector* (2021) and the National Academies of Sciences, Engineering, and Medicine in *Accelerating Decarbonization of the U.S. Energy System: Technology, Policy, and Societal Dimensions* (2021). These two significant reports indicate that a rapid transition is possible, but it will require major governmental policy efforts and societal change—market dynamics alone cannot create the shift. Although many improvements can happen with existing technologies, others require innovation to bring critical technologies and energy resources to market at the scale of production that is needed. Within the energy system, three strategies are key: (1) reducing the overall demand for energy, (2) decarbonizing energy fuels, and (3) converting as many energy technologies as possible to electricity, such as conventional cars to electric vehicles.

The U.S. Energy Information Agency believes that there will be a continued shift from coal and other fossil fuels to a national energy portfolio that has an increasing percentage of renewable energy, as shown in Figure 7.2. Many countries have set ambitious goals to convert to 100% renewable energy by 2050. Renewable energy has seen rapid and accelerating growth in recent years, including a 45% annual rate of increase of global renewable energy capacity in 2020, according to the International Energy Agency. In 2021, renewable energy growth in the United States was greater than for both coal and nuclear energy. On October 7, 2022, Greece reported running on completely renewable energy for several hours. It is of particular interest to note that CO_2 is removed through natural processes over time. As such, there is no requirement to eliminate all fossil fuels, but there is an opportunity to plan for a future that society wants, balancing renewable energy and supplementing renewable energy systems with fossil fuels while considering the economic, social, and environmental impacts from fossil fuel use.

The world's dependence on fossil fuels has created a host of environmental problems, and as finite resources, they cannot be used indefinitely into the future. Our challenge today is to provide for the unmet needs of the world's energy poor and to maintain safe, reliable, abundant, and affordable energy systems for all. To accomplish this goal, we must engineer or reengineer our energy supply in a variety of ways. First, we can diversify our resources by substituting cleaner and more renewable forms of energy for electrical, mechanical, and thermal energy. Second, we can enhance energy conservation by making our systems more efficient, by recovering otherwise wasted energy, and by inventing devices and materials that curtail the amount of energy required. Third, we can rethink the scale and distribution of our energy technologies to take better advantage of small-scale energy generation.

Implications and Simulation of Energy Choices and Climate Change

In this activity, you will use the En-ROADS simulation to explore the consequences of energy policy decisions, economic growth, land use choices, and other policies and their impacts on global warming. This simulation was developed by Climate Interactive, Ventana Systems, the UML Climate Change Initiative, and the MIT Sloan Business School to show the relationships between energy choices and climate change (Chikofsky et al., 2022). You will use the En-ROADS simulation model to create a set of policies that may mitigate global warming to the internationally agreed target of less than 2°C (3.6°F) warming over the surface of the planet.

The first step is to get acquainted with the En-ROADS program. You can access En-ROADS online at en-roads.climateinteractive.org/scenario.html?v=22.10.0. The En-ROADS interface, shown in Figure 7.5, allows changes via the interactive slider tools to the energy source portfolio, transportation type and efficiency, building efficiency, economic growth, and changes in land use. En-ROADS also allows you to consider carbon removal technologies. The En-ROADS one-page guide to the control panel, shown in Figure 7.6, provides definitions of each variable category.

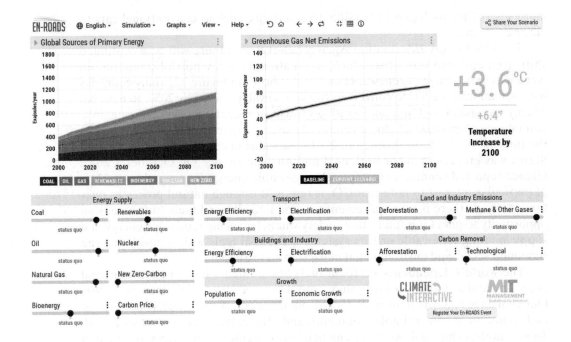

Figure 7.5 The control panel for the En-ROADS energy and climate simulation developed by Climate Interactive, Ventana Systems, the UML Climate Change Initiative, and the MIT Sloan Business School.
Source: en-roads.climateinteractive.org/scenario.html?v=22.10.0.

7.7 Create a scenario to demonstrate an energy development plan using the En-ROADS simulator that limits global warming to less than 2°C (3.6°F) as agreed in the Paris Accord. Note: There are many pathways to achieving the Paris goals to limit climate change, so you may focus on what you believe is most appropriate. Adjust the scenario and provide a brief explanation for each of the scenario variables below.

a) How do you believe the *population in the United States, the European Union, China, India, and developing countries* will change across the planet by 2100? Population projections can be found from a variety of sources online. Be sure to document the source you use. Adjust the sliders on the simulator to account for the impact of population projections.

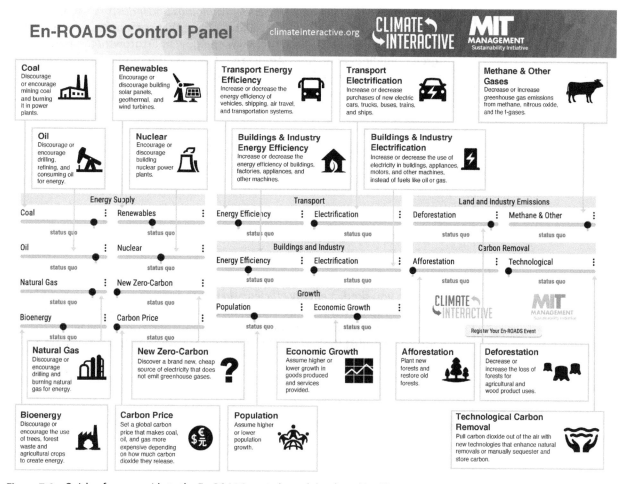

Figure 7.6 Quick reference guide to the En-ROADS control panel developed by Climate Interactive, Ventana Systems, the UML Climate Change Initiative, and the MIT Sloan Business School.
Source: www.climateinteractive.org/wp-content/uploads/2019/09/EnROADS-one-page-guide-to-control-panel-v11-dec-2021.pdf.

b) How do you predict *economic growth by gross domestic product per capita in the United States, the European Union, China, India, and developing countries* will change by 2100? Economic growth projections can be found from a variety of sources online. Be sure to document the source you use. Adjust the sliders on the simulator to account for the impact of population projections.

c) How do you predict *energy efficiency in buildings and industry* will change by 2100? The U.S. EPA has a table available online that compares Green Building Standards that might be helpful (www.epa.gov/smartgrowth/comparison-green-building-standards). Adjust the sliders on the simulator to account for the impact of population projections.

d) How do you predict *energy efficiency in transportation* will change by 2100? The U.S. Department of Transportation and other organizations may help you estimate reasonable improvements in transportation between 2020 and 2026. Adjust the sliders on the simulator to account for the impact of population projections.

e) How do you predict *the sources for energy supply* will change by 2100? The U.S. Energy Information Agency's Annual Energy Outlook (www.eia.gov/outlooks/aeo) provides a great deal of insight into how energy supplies may change between the present and 2050. Adjust the sliders on the simulator to account for the impact of population projections.

f) How do you predict *the sources for land and industry emissions* will change by 2100? Adjust the sliders on the simulator to account for the impact of population projections.

g) How do you predict *carbon removal* will change by 2100? Adjust the sliders on the simulator to account for the impact of population projections

7.8 Create a written or video report to demonstrate an energy development scenario using the En-ROADS simulator that limits global warming to less than 2°C (3.6°F) as agreed in the Paris Accord. Include each of the items below in your report:

a) Describe how you predict *changes in population in the United States, the European Union, China, India, and developing countries* by 2100. Adjust the sliders on the simulator to account for all your scenario projections. Illustrate your projection impacts on population using the graphs to illustrate the *population by region* (on the left-hand graph) and the *population exposed to sea-level rise* (on the right-hand graph). Take a screenshot of these graphs once all your scenario sliders have been finalized and include your visual and the interpretation of the graphs in your report.

b) Describe how you predict *changes in economic growth in the United States, the European Union, China, India, and developing countries* by 2100. Adjust the sliders on the simulator to account for all your scenario projections. Illustrate your projection impacts on economic growth using the graphs to illustrate the *gross domestic product per capita by region* (on the left-hand graph) and the *marginal cost of solar electricity history* (on the right-hand graph). Take a screenshot of these graphs once all your scenario sliders have been finalized and include your visual and the interpretation of the graphs in your report.

c) Describe how you predict *changes in building and industrial efficiency* by 2100. Illustrate your projection using the graphs to illustrate the *electric share of total capital—buildings and industry* and the *marginal cost of solar greenhouse gas net emissions history*. Take a screenshot of these graphs once all your scenario sliders have been finalized and include your visual and the interpretation of the graphs in your report.

d) Describe how you predict *changes in energy efficiency in transportation* by 2100. Illustrate your projection using the graphs to illustrate the *electric share of total capital—transport* and the *air pollution from energy by source (area)—PM2.5*. Take a screenshot of these graphs once all your scenario sliders have been finalized and include your visual and the interpretation of the graphs in your report.

e) Describe how you predict *changes in the sources for energy supply* by 2100. Illustrate your projection using the graphs to illustrate the *global sources of primary energy (area)* and the CO_2 *concentration*. Take a screenshot of these graphs once all your scenario sliders have been finalized and include your visual and the interpretation of the graphs in your report.

f) Describe how you predict *changes in the sources for land and industry emissions* by 2100. Illustrate your projection using the graphs to illustrate the *greenhouse gas net emissions (area)* and the *greenhouse gas concentration*. Take a screenshot of these graphs once all your scenario sliders have been finalized and include your visual and the interpretation of the graphs in your report.

g) Describe how you predict *changes in carbon removal* by 2100. Illustrate your projection using the graphs to illustrate the *CO_2 emissions and removal* and the *temperature change*. Take a screenshot of these graphs once all your scenario sliders have been finalized and include your visual and the interpretation of the graphs in your report.

h) **What is the greatest financial challenge or uncertainty toward achieving your energy scenario? Illustrate your projection using the graphs to illustrate the** *financial* **challenge. Take a screenshot of the graph once all your scenario sliders have been finalized and include your visual and the interpretation of the graph in your report.**

i) **What is the greatest environmental challenge, uncertainty, or impact in your energy scenario? Illustrate your projection impacts using the simulation graphs to illustrate the** *biological or environmental* **challenges. Take a screenshot of the graph once all your scenario sliders have been finalized and include your visual and the interpretation of the graph in your report.**

j) What is the greatest challenge to human health and well-being in your energy scenario? Illustrate your projection impacts using the simulation graphs to illustrate the *humanitarian* challenges. Do you believe people living in high-income and low-income countries will be equally affected by energy use scenarios in the future? Take a screenshot of the graph once all your scenario sliders have been finalized and include your visual and the interpretation of the graph in your report.

k) What steps need to be taken (include science development, engineering, technology transfer, education, political action, policy, and so on) over the span of the next five years to put your plan into action?

References

Chikofsky, J., Johnston, E., Jones, A., Zahar, Y., Campbell, C., Sterman, J., Siegel, L., Ceballos, C., Franck, T., Kapmeier, F., McCauley, S., Niles, R., Reed, C., Rooney-Varga, J., and Sawin, E. (2022). *En-ROADS User Guide*. Developed by Climate Interactive, Ventana Systems, the UML Climate Change Initiative, and the MIT Sloan Business School. April 2022. docs.climateinteractive.org/projects/en-roads/en/latest/index.html, accessed October 12, 2022.

Hitaj, C., and Suttles, S. (2016). *Trends in U.S. Agriculture's Consumption and Production of Energy: Renewable Power, Shale Energy, and Cellulosic Biomass*, EIB-159. Washington, DC: U.S. Department of Agriculture, Economic Research Service, August.

International Energy Agency. (2021). *Renewable Energy Market Update 2021*. Paris: International Energy Agency. www.iea.org/reports/renewable-energy-market-update-2021, accessed October 14, 2022.

National Academies of Sciences, Engineering, and Medicine. (2021). *Accelerating Decarbonization of the U.S. Energy System: Technology, Policy, and Societal Dimensions*. Washington, DC: The National Academies Press, p. 268. doi.org/10.17226/25932

Suits, R., Matteson, N., and Moyer, E. (2020). "Energy Transitions in U.S. History: 1800–2019." US Energy History Visualization. us-sankey.rcc.uchicago.edu, accessed November 16, 2022.

U.S. Energy Information Administration. (2022). "Annual Energy Outlook 2022 (AEO 2022)." www.eia.gov/outlooks/aeo/pdf/AEO2022_ChartLibrary_full.pdf.

U.S. Environmental Protection Agency. (2022). *Clean Energy Calculations and References*. www.epa.gov/cleanenergy/energy-resources/refs.html, accessed October 12, 2022.

Life Cycle Thinking: Understanding the Complexity of Sustainability

Objectives

Chapter 8 focuses on understanding the environmental impacts of complex interactions related to choices in the design process. Students will develop basic knowledge and tools to consider the life cycle of a product during the design phase. Life cycle thinking allows us to compare the impacts associated with different products or processes to determine which have the least impact. It may also allow us to determine which stage in the overall life cycle of products contributes the most to the overall impact.

Student Learning Objectives

- Understand the role of life cycle thinking in sustainable design.
- Describe the components and steps utilized to develop a life cycle analysis for a product.
- Appreciate the limitations of life cycle analysis.

During the design process, one goal may be to minimize the environmental impacts of a product. This goal requires comprehensive tracking of the way materials and energy flow through the product system—from the extraction of virgin raw material from the environment through the various stages of processing, refinement, packaging, sales, and use all the way to the ultimate disposal back into the environment. In this chapter, we will use a fly-fishing rod, a product made of only a few components, to consider how the energy and material flows impact the sustainability of the product.

8.1   Introduction

Industrial engineers focus on creating efficient manufacturing and production processes. Industrial and systems engineers work across a variety of industries, including automobile manufacturing and aerospace, health care, forestry, finance, leisure, and education. They devise efficient systems that integrate workers, machines, materials, information, and energy to make a product or provide a service. Typically, responsibilities of industrial engineers include managing production schedules and process flows, ensuring quality control, and designing control systems to ensure that products meet quality standards.

The roots of industrial engineering began with the industrial revolution and the mechanization of previously manual operations in the textile industry. Industrial engineers worked in manufacturing plants and were involved with the operating efficiency of workers and machines. Today, they design manufacturing systems to optimize the use of computer networks, robots, and materials. Industrial engineers develop a versatile skill set that allows them to engage in activities such as supply chain management, quality assurance, and project management. Industrial engineers may also be called manufacturing engineers and systems engineers, depending on their responsibilities. With a focus on management, efficiency, and waste reduction, industrial engineers are well positioned to help discover and implement new, more sustainable ways to produce and manufacture products. We will examine how industrial and systems engineers consider a life cycle approach to design and make product improvements by studying how materials are turned from raw materials into a product; in this case study, the product will be a fly-fishing rod.

Fly-fishing is an activity that may be done in cold-water streams, warm-water rivers, estuaries, and oceans. It is an activity that is possible for all ages and has been shown to have physical and emotional benefits. Because fly-fishing requires relatively clean water and natural habitat for the fish, many fly-fishing manufacturers and companies promote the idea of sustainability and environmental conservation. However, promoting sustainability may be much easier than producing sustainable products and systems. In this chapter, we will explore how the entire life cycle of a product plays a role in the analysis of the sustainability of that product. In order to explore this possibility, we will first note your preconceived notions about the product and sustainability and examine how they might change throughout the process of applying life cycle thinking to the design and manufacturing of products.

8.1 Describe how the U.S. Bureau of Labor and Statistics defines an industrial engineer, the entry-level education required, and the median salary for industrial engineers by reviewing the Field of Degree highlights for students and job seekers in the *Occupational Outlook Handbook*.

8.2 Describe how the U.S. Bureau of Labor and Statistics defines a materials engineer, the entry-level education required, and the median salary for a materials engineer by reviewing the Field of Degree highlights for students and job seekers in the *Occupational Outlook Handbook*.

8.3 Find job listings for an industrial engineering position and a materials engineering position. Describe the similarities and difference between the education required, experience required, and salary range (if available) for the two different positions. What other educational degrees or requirements would be acceptable for the job listings you have found? You can find job listings through employment websites, USAJobs.gov, IISE.org, LinkedIn, and other professional networking websites.

Part 1: Preconceptions about Green Design

You or your design team will work to design the "greenest," most "environmentally friendly" fly-fishing rod. The design will be judged for its intended use and sustainability. The objective of this part of the exercise is to capture your initial thoughts and concepts about your design and how you might utilize sustainability as a parameter in the design process. This exercise is a preliminary assessment. You *should not* use any additional resources. There are no wrong answers.

Describe your idealized fly-fishing rod below.

8.4 Describe the weight of the rod.

8.5 Describe the length of the rod.

8.6 Create a list of materials you think you may need to make the rod.

8.7 Describe how you will judge if your rod is more sustainable (greener, more environmentally friendly) than other teams. Create a table that shows the following:

a) Five or more things you might measure to describe how your design is more or less sustainable than other designs.

b) List the unit of measure for each of the metrics you list in part a.

8.2   How a Fly-Fishing Rod Works

The purpose of a fly-fishing rod is quite difficult to achieve: its role is to throw or cast a small bunch of feathers on a small hook 5 to 25 m toward a moving fish! Modern fly-fishing rods have evolved using both traditional and modern materials to achieve this purpose. The fly-fishing rod performs the role of two simple machines: a lever and a spring. A fly-fishing rod is quite long, usually between 2 to 3 m; this way, it gains the full advantage of the lever action. The rod acts as an extension of the arm, as illustrated in Figure 8.1. The long length allows additional force to be added to the fly-fishing line as it is cast through the air. In addition, the rod is designed to flex or load due to the mass and momentum of the moving fly-fishing line. Once loaded or reaching the maximum curvature, as shown in Figure 8.1, the rod will spring back in the forward direction, imparting additional force to propel the fly-fishing line forward.

The act of casting a fly rod involves timing and practice to anticipate how the rod acts as a lever and spring. A fly cast involves a back cast, using a backward motion of the fly rod, and a forward cast, utilizing a forward motion of the fly rod. The fly cast begins with a backward motion of the rod to lift the fly line from the

Casting Phases

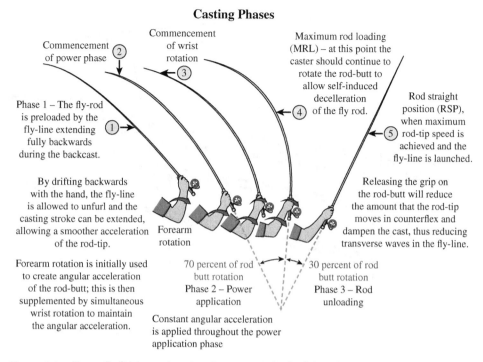

Commencement of power phase ②

Commencement of wrist rotation

③

Maximum rod loading (MRL) – at this point the caster should continue to rotate the rod-butt to allow self-induced deceleration ④ of the fly rod.

Phase 1 – The fly-rod is preloaded by the fly-line extending fully backwards during the backcast. ①

Rod straight position (RSP), when maximum rod-tip speed is achieved and the fly-line is launched. ⑤

By drifting backwards with the hand, the fly-line is allowed to unfurl and the casting stroke can be extended, allowing a smoother acceleration of the rod-tip. Forearm rotation

Releasing the grip on the rod-butt will reduce the amount that the rod-tip moves in counterflex and dampen the cast, thus reducing transverse waves in the fly-line.

Forearm rotation is initially used to create angular acceleration of the rod-butt; this is then supplemented by simultaneous wrist rotation to maintain the angular acceleration.

70 percent of rod butt rotation
Phase 2 – Power application

30 percent of rod butt rotation
Phase 3 – Rod unloading

Constant angular acceleration is applied throughout the power application phase

Figure 8.1 How a fly-fishing rod works when casting the fly-fishing line.

water and impart a rearward velocity to the line to load the spring action of the rod. Subtle motions are all that are required to utilize the level action of the fly rod. However, the motions must be precise since the rod becomes essentially a 3-m extension of each motion of your wrist, elbow, and forearm. Practice is needed for developing the timing related to how the spring function of the rod is loaded with the mass and momentum of the line. To maximize the distance of the cast, the rod should pause briefly after the back cast. This brief pause loads the rod in order to maximize the spring force that will help propel the line forward after the forward cast begins. When casting, you can feel the force of the fly line loading the rod. Once the fly line's momentum is transferred to the rod, the forward cast begins, and the line is accelerated in the forward direction from the spring action of the fly rod. Careful efficient motions backward and forward help to provide the maximum transfer of force from the line to the rod (spring). Careful timing of the back cast and forward cast allow for maximum benefit of the spring action to provide the maximum distance the line can be cast after the forward cast. There are excellent videos available online that explain this motion and show the action of the fly rod and fly line.

The mass, action, and length of the fly rod vary greatly according to the type of fish species you are targeting, the expected mass of the fly used for that species, and personal choice.

A fly rod like that shown in Figure 8.2 is a mechanical tool made of a rod blank, a handle, a reel seat, fly-line guides, a wrapping material to hold the guides in place, and a protective coating over the rod or parts.

The fly rod **blank** is the first choice of component to select when building a fly rod, as all the other parts are designed to work with the rod blank. The blank is the part of the rod that serves as the backbone of the entire rod. The blank may be 2 to 3 m in length. The length of the rod is often chosen based on the type of water and

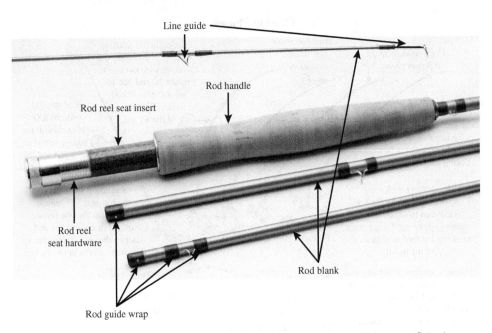

Figure 8.2   Components of a 5-weight, 2.75-m-long, four-piece graphite composite fly rod.

the characteristics of the material of the rod blank. Anglers fishing in small streams that are covered by small trees and brush may prefer rods from 1.8 to 2.5 m. Anglers fishing along an ocean beach or large river might prefer a rod that is up to 3.6 m long. The most common rod blank length is 2.75 m, a very good compromise between the shorter and longer lengths.

The **weight** of a fly-fishing rod is not a description of the mass of the rod. Instead, the weight of the fly rod refers to the mass of the fly-fishing line it is designed to cast. A heavier-weight rod will cast a fly line with a greater mass. For example, a 3-weight rod is designed for small streams, and a low-mass fly line is designed to cast tiny flies to relatively small fish, as illustrated in Figure 8.3. The lightest weight rod blanks, 1- to 3-weight blanks, are used for smaller species of fish and when lightweight fly line is required because smaller trout and panfish may be very sensitive to vibrations in the water that can be caused by heavier-weight fly lines. A 4- to 7-weight fly rod is designed to cast artificial flies for midsized freshwater fish such as bass and trout. These fish require a heavier-weighted 4- to 7-weight fly line and rod blank for longer casts and the ability to fight the fish and bring them into the fishing net. Someone fishing for a large saltwater species such as tarpon will use a very large "fly," sometimes over 8 inches in length, which is quite heavy. Anglers often prefer the heaviest-weight 8- to 14-weight lines and fly rod blanks for making long casts in salt water and bringing large fish species to the landing net. Tarpon fishing requires a strong and heavy fly-fishing line and may use a 12-weight fly line and rod. The fly rod blank is therefore chosen to match the type of fish species being targeted. A larger-numbered "weight" fly rod will usually have a thicker diameter for the strength required to reel in the targeted fish species. Generally, the wider the diameter of the blank, the greater the mass. The diameter at the base of a fly rod may be 0.4 to 1.3 cm, narrowing to less than 1 mm at the tip of the blank for some lighter-weight rods.

In addition to the length and weight of the fly rod blank, the flexibility of the blank is also a major consideration of the angler when choosing a fly rod. The

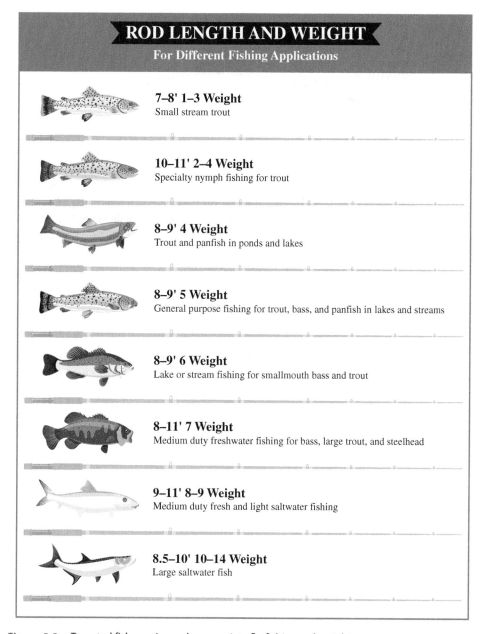

ROD LENGTH AND WEIGHT
For Different Fishing Applications

7–8' 1–3 Weight
Small stream trout

10–11' 2–4 Weight
Specialty nymph fishing for trout

8–9' 4 Weight
Trout and panfish in ponds and lakes

8–9' 5 Weight
General purpose fishing for trout, bass, and panfish in lakes and streams

8–9' 6 Weight
Lake or stream fishing for smallmouth bass and trout

8–11' 7 Weight
Medium duty freshwater fishing for bass, large trout, and steelhead

9–11' 8–9 Weight
Medium duty fresh and light saltwater fishing

8.5–10' 10–14 Weight
Large saltwater fish

Figure 8.3 Targeted fish species and appropriate fly-fishing rod weight.

amount of flexibility in the rod is called the rod's **action**. A more flexible rod blank that corresponds to a spring that deforms very easily is described as having a slow action, as illustrated in Figure 8.4. A less flexible rod blank that corresponds to a strong spring that requires more force for deformation is described as having a fast action. Rod blanks with intermediate flexibility have a medium action. Typically, a beginner may want more time and greater ability to feel the spring force action of the rod blank and may prefer a slower-action rod blank. Anglers with more experience casting and those using heavier flies that impart a greater force on the rod may prefer a faster-action rod blank. The flexibility of the rod is related to the characteristics of the material from which the fly rod blank is made.

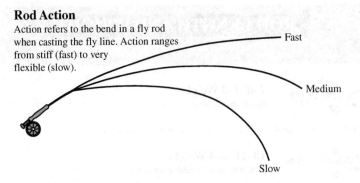

Rod Action
Action refers to the bend in a fly rod when casting the fly line. Action ranges from stiff (fast) to very flexible (slow).

Fast

Medium

Slow

Figure 8.4 The flexibility of the rod blank is described as the rod's action, which may be a slow action (most flexible, weakest spring action), medium action, or fast action (least flexible, strongest spring action).

Part 2: Choosing the Type of Fly-Fishing Rod for the Customer

8.8 Briefly describe the mechanics of the forces that act on the fly rod. You may also use resources (there are many) that describe the physics of fly casting. Use a sketch to show the forces acting on the rod during the following:

a) The back cast
b) The pause between the back cast and the forward cast
c) The forward cast

8.9 Briefly describe the casting process and the *role of the fly-fishing rod* in that process. The Fly Rod Selector may help you get a sense of important questions to ask when designing your rod (www.orvis.com/fly-rod-selector.html). You may also find information from a local fly-fishing shop if there is one located nearby.

8.10 Briefly describe the *role of the fly-fishing line* in the casting process, including a description of what type of line might be most appropriate for your intended use.

8.3   Design Choices for a Fly Rod

Three materials—bamboo, fiberglass, and graphite composites—are used in fly-fishing rods today. When designing and building a fly rod, the first decisions to make involve the weight, length, and action of the fly rod that are suitable for the location and type of fish species being targeted. The weight, length, and action help the designer determine the type of material that is best suited to the type of fishing experience desired. Each material has different characteristics that relate to the action of the blank and have certain advantages for specific uses.

Historically, the technique of fly fishing can be dated back as early as AD 200 in Macedonia. The first fly-fishing rods were made of wood materials that provided the ability to flex and recover. As fishermen searched for ways to improve the design of the wood rods, they developed better materials and technologies. In the 17th century, anglers began developing hollow wooden tubes to create specialized rods. Bamboo is a fibrous evergreen grass that develops a flexible hollow woody cellulose structure called a cane as it grows. The hollow wood tubes were used to make fishing "poles." In the 18th century, rod makers realized that the strength and flexibility of the rods could be greatly increased and the weight significantly decreased by creating bamboo strips that could be joined together to make an effective fly rod. Fly rod makers settled on one particular type of bamboo cane, *Pseudosasa amabilis*, for making fly rods that is grown specifically in the Tonkin region of China. These bamboo canes are extremely straight, thick walled, and have smaller, non-prominent nodes. The bamboo cane is dried and then split into small tapering triangular sections that are glued and lashed together in an octagonal pattern to create a solid strong rod blank, as shown in Figure 8.5. The bamboo rod blank was the basis of most fly-fishing rods for a period of almost 200 years! The bamboo rod blanks are very flexible and resilient natural materials that have been used for over a

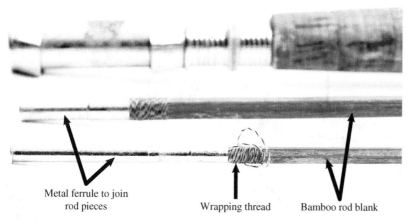

Metal ferrule to join
rod pieces

Wrapping thread

Bamboo rod blank

Figure 8.5 A bamboo fly rod and components.

century to produce some of the most sought-after fly rods. However, bamboo rod blanks are heavier than other modern materials, and the process of making the rod blanks is labor and time intensive.

Fiberglass is a glass-reinforced composite material utilizing glass fibers for strength and rigidity and an epoxy, polyester resin, or vinyl ester resin as the thermoplastic material. In a composite material, the fibers carry most of the load imparted by the force, and the type and alignment of the fibers control the mechanical properties of the material. The resin helps to transfer load between fibers, prevents the fibers from buckling, and binds the materials together. Large-scale production of fiberglass fly rods began in the 1950s, and fiberglass rods (see Figure 8.6) became more common than bamboo rod throughout the 1970s. Fiberglass materials are very strong, especially compared to the mass of the material used. These composite materials are said to have a high strength-to-weight ratio. The relatively long fiberglass composite fibers simulate the natural bamboo fibers, and the way the rod flexes when cast with a fly line is like the slow-bending action anglers were accustomed to with traditional bamboo rods. However, fiberglass rods can be mass-produced and require far less labor in manufacturing. Fly rod

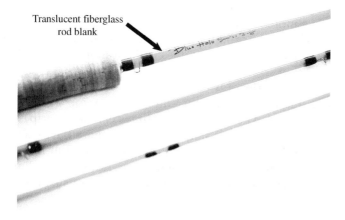

Translucent fiberglass
rod blank

Figure 8.6 A 3-weight, 2.3-m fiberglass fly rod.

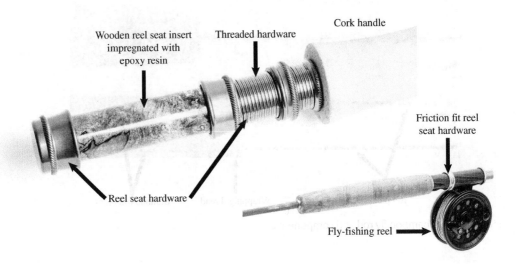

Wooden reel seat insert
impregnated with
epoxy resin

Threaded hardware

Cork handle

Friction fit reel
seat hardware

Reel seat hardware

Fly-fishing reel

Figure 8.7   The reel seat insert and hardware used to hold the reel onto the fly rod.

manufactures make fiberglass fly rods for their slow to medium action, moderate weight, overall durability, and strength.

Graphite composite materials (sometimes called carbon composites) are composed of organic polymer fibers such as polyacrylonitrile and a resin. The graphite fibers are kept under tension and are heated under high temperatures (>1,000°C) to create carbon-carbon crystals (graphite). The carbon fibers have extremely strong molecular bonds that produce a very high strength-to-weight ratio. Graphite composite fly rods were introduced in the 1970s and have become the most popular type of fly rod sold today. Today's graphite rods, like that shown in Figure 8.2, can be produced at the same price point as fiberglass rods, as the production process is similar. The graphite rods can be narrower in diameter and longer in length and have less mass than fiberglass rods. Because the graphite composite materials provide greater rigidity than fiberglass materials, graphite rods have a medium to fast action. For advanced flycasters, the fast action can produce a stronger spring imparted force. If the angler's motion and timing are correct, this can result in a more efficient transfer of forces between the angler's motion and the fly rod, which may produce greater casting speed and length. However, the graphite fly rod's greater rigidity does make it more prone to breaking than fiberglass rods.

The rod blank is not the only component of the fly rod. All the components of the rod, including the handle, reel seat, guides, guide wrap, and protective coating, must be chosen. The handle (sometimes called the grip) of the fly rod, shown in close-up in Figure 8.7, is simply the place where the caster's hand holds the rod. The fly rod blank passes through the hollowed-out handle. Any comfortable material can be used for the handle; cork and foam handles are the most common materials used today, but perhaps more sustainable materials might be created from recycled products in the future. The very bottom of the fly rod blank passes through the handle and the reel seat. The reel seat holds the fly-fishing reel to the fly rod. The reel seat is composed of a round hollow reel seat insert against which the reel rests on the rod. The insert can be made of any material that is rigid and durable and can be shaped to the form required. The real seat hardware may consist of friction fitted rings or shaped rings and threaded nuts that hold the reel onto the insert and rod, as shown in Figures 8.1 and 8.7. Reel seats are typically made of aluminum or steel for strength and durability.

The guides on a fly-fishing rod do exactly as the name implies: they guide the fly line along the length of the fly rod. The guides, shown in Figures 8.2 and 8.8, are positioned along the rod blank. There are a variety of types of guides available, most of which are made of aluminum or stainless steel. Some guides may have ceramic, plastic, nylon, or stone inserts to provide a smooth surface for the line to run through. The guides are most often held in place by thread wrapped tightly over a portion of the guide called the foot, which is attached to the rod blank and held in place by the wrapped thread. The 75-year-old bamboo rod shown in Figure 8.5 shows what happens if the wrapped thread begins to unravel over time. This old bamboo rod was coated with a shellac to protect the rod blank and wraps. Many different types of protective coating can be used to protect the rod blank and the line guides. Most fiberglass and graphite rods use epoxy resin coatings that are applied over the thread to protect the guide wraps. Natural epoxies and coatings, as well as synthetic epoxies and resins, are still used by rod manufacturers to protect the finished rod, with each having aesthetic, environmental, and functional advantages and disadvantages.

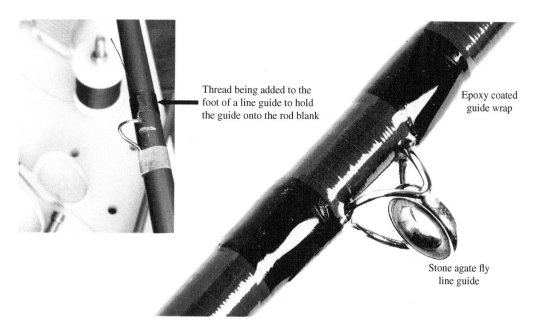

Thread being added to the foot of a line guide to hold the guide onto the rod blank

Epoxy coated guide wrap

Stone agate fly line guide

Figure 8.8   Guide wraps and protective coatings.

Part 3: Materials and Design

8.11 Understanding the material choices is critical in design and manufacturing. Often, there may be many alternative materials that function appropriately for the design problem. There will be trade-offs in terms of cost, function, and environmental impacts. Briefly describe the materials you choose for each of the components listed.

a) Rod blank
b) Handle
c) Reel seat insert and hardware
d) Fly line guides
e) Guide wrapping thread
f) Protective coating to cover the guide wraps

8.12 Cite one or more sources (website, catalog, store, and so on) for the materials chosen. Note: The sources may be updated or changed later in the project.

8.13 What is the cost of each component used in the design of the fly rod?

8.14 Describe the *raw* materials associated with each component. Find a possible geographic location and repository for the raw materials of each component (i.e., where in the world are the materials found, mined, or grown).

8.4 Materials Flow Analysis

Materials flow analysis is a method that can be used to develop the inventory for a life cycle analysis. It is a systematic assessment of the flows of materials within the boundaries of a system in space and time. Materials flow analysis relies on the law of conservation of mass and is performed by simple mass balances comparing inputs, outputs, depletions, and accumulations within systems and subsystems. Some relevant terminologies that are used in materials flow analysis are the following:

- **Materials:** Substances and goods.
- **Process:** The transport, transformation, or storage of materials.
- **Reservoir:** A system that holds materials. It can be a source from which materials come, a sink into which materials go, or both.
- **Stocks:** The quantity of materials held in reservoirs within a system.
- **Flows/Fluxes:** The ratio of the mass per time that passes through a system boundary is termed the *flow*, while the *flux* is the flow per cross section.

One of the applications of mass flow analysis in industrial systems is the ability to track how much of the original materials going into the process end up as part of the desired product stream. Applying the principles of conservation of mass, we know that only a portion of the raw materials entering the system becomes the product, while the rest ends up in the waste streams. The efficiency of the process is defined as the ratio of the mass in the desirable product and the total mass entering the system. For a 100% efficient system, no emissions are generated, and all the raw material is converted into useful product. This is probably not a realistic goal for industrial systems; a more feasible aspiration is for industrial systems to mimic a process in which limited waste is generated across the entire industrial system. The most efficient system will be the one in which the total waste emitted from the system is minimal. The *reuse factor* is defined as the amount of recycled material that can be recovered and used in the process to make a product.

The boundaries of a materials flow system can be drawn across different spatial levels, from a single industrial facility to national and even global levels. All materials flow analyses entail the construction of a materials budget, which requires the following:

1. Determining the material(s) to be evaluated.
2. Identifying all relevant reservoirs and flow streams.
3. Quantifying the contents of the reservoir and the magnitudes of the flows. This information is obtained from data if available, taking measurements where possible, and, finally, making estimations if necessary.

8.15 Create a *draft sketch* of the material flow process for making a fly rod. Start with the raw material required (cradle). Include the intermediate steps required to process the raw materials and create engineered materials used to make each of the components of the fly rod. Don't forget the transportation steps that are required to move the material from one step to the next in your sketch. Finally, consider the processes required to get the fly rod to the customer. Note: This will be a draft and will be updated as we move forward with the project and design.

8.5 Energy Flow Analysis

Energy is the ability to do work, and it is an essential need for a functional industrial system. The objective of sustainability is to reduce and minimize the energy requirements of product manufacture. Energy flows in industrial systems are also tracked closely with materials flows because they are often intrinsically linked.

One way in which energy flows are assessed is through the concept of energy efficiency. Whenever energy from any source is converted into work, some of the energy input is dissipated to the surroundings. The energy of this system is defined as the ratio of the useful work obtained and the total energy input. For example, approximately 25% of the energy derived from combustion of gasoline is used to propel a car; the rest is dissipated to the surroundings as heat. Thus, the efficiency of an average automobile is 0.25.

Like the concept of embodied water, the total amount of energy required in producing a product or service across the product chain is called the embodied energy. A significant amount of energy used to manufacture and use products is hidden and not always accounted for by manufacturers and consumers.

Part 4: Materials and Energy Flow Visualization

8.16 For the components chosen below, create a graphic to visualize the flow of materials and energy throughout the design life of a fishing rod. Each member of your team may choose a different component. The schematic should include manufacturing steps in acquiring the raw material (cradle), processing the raw material, product manufacturing, product sales, product use, and product disposal (grave), as well as the transportation steps and types of fuel consumed in each. This is a qualitative visual tool; it does not need to be quantitative.

a) Rod blank
b) Handle
c) Reel seat insert and hardware
d) Fly line guides
e) Guide wrapping thread
f) Protective coating to cover the guide wraps

8.17 For the components chosen, add elements that include the flow of energy throughout the cradle-to-grave life cycle of a fishing rod. Each member of your team may choose a different component. Embodied energy is fundamentally a value that assumes that the production processes for any given material are identical. Is this a valid assumption? Why or why not?

8.18 Identify on the diagrams from Problems 8.16 and 8.17 the sources of waste energy or materials. For example, this could be wasted heat in electricity production or wasted cork powder created when manufacturing the handle.

8.19 Tables for embodied energy for various materials are provided in Table 8.1. Using the table or other referenced sources, determine the embodied energy for the individual components (a–f). Note that this will include determining the mass associated with the various components. The component weights may be determined through vendor supplied information.

8.20 Calculate a *sustainability indicator* based on embodied energy for your design. Using the embodied energy data from your group, calculate the total embodied energy for your team's fishing rod design. Hint: If this is done in a spreadsheet, it can be easily updated in case your design changes in the future.

Table 8.1   Embodied energy of materials

Material	Embodied Energy (MJ/kg)	Material	Embodied Energy (MJ/kg)
Aluminum—general[1]	155	Plastic—polypropylene[1]	95.4
Aluminum—recycled[1]	29	Plastic—polystyrene[1]	86.4
Aluminum—virgin[1]	218	Plastic—polyurethane[1]	72.1
Brass[1]	44	Polyester fabric[2]	53.7
Carbon (graphite) fiber composite[3]	315	PVC[1]	77.2
Carpet—nylon[1]	130	Resin—epoxy[1]	137.0
Ceramics[1]	10.0	Resin—melamine[1]	97.0
Copper—tube and sheet[1]	42.0	Rubber[1]	91.0
Cork insulation[1]	4.0	Silver[1]	128.2
Cotton fabric[1]	143	Steel—general[1]	26.2
Glass[1]	15.0	Steel-galvanized sheet—recycled[1]	13.1
Glass-reinforced plastic (GRP) fiberglass[1]	100	Steel-galvanized sheet—virgin[1]	40.0
Paint—general[1]	70	Steel—recycled wire[2]	12.5
Paint—acrylic[2]	61.5	Steel—stainless[1]	56.7
Paint—solventborne[1]	97	Urea sealant[2]	78.2
Paint—waterborne[1]	59	Water[1]	0.01
Paint—wood stain/varnish[1]	50.0	Wax[1]	52.0
Plastics—general[1]	80.5	Wood—air-dried hardwood[2]	0.5
Plastic—ABS[1]	95.3	Wood—glue-laminated timber1	12.0
Plastic—HDPE[1]	76.7	Wood—kiln-dried softwood[2]	3.4
Plastic—LDPE[1]	78.1	Wood—kiln-dried hardwood[2]	2.0
Plastic—polycarbonate[1]	112.9	Wood—laminated veneer lumber[1]	9.5
Plastic—polyester[2]	53.7	Wood—particleboard[1]	14.5

Sources: Hammond and Jones (2011)[1]; Lawson (1996)[2]; Kara and Manmek (2009)[3].

8.6   Life Cycle Thinking

The life cycle of a product starts when the raw materials are collected to create a product and "ends" when the product is disposed of or repurposed. When an engineer looks to reduce the waste in the production of a product or create a more sustainable version of a product, the entire life cycle should be considered. *Life cycle thinking* expands the focus of the engineering from a singular focus on one step, typically the manufacturing processes, to include the processes and uses of the product materials over their entire life cycle. The producer becomes responsible for the products from cradle to grave and must, for example, develop products with improved performance in all phases. The main goal of life cycle thinking is to improve positive impacts of the product from an economic and social aspect while reducing negative economic, environmental, and social impacts of the product throughout the stages of the product's life.

The life cycle of a product or service is defined by the International Organization for Standardization (ISO) as "consecutive and interlinked stages of a product system, from raw material acquisition or generation from natural resources to final disposal" (ISO 14040). Life cycle thinking can be quantified by measuring, modeling, or estimating impacts associated with each step of the life cycle through the process called *life cycle assessment* (LCA). The specific processes of ISO 14040 and 14044 are defined as "the compilation and evaluation of the inputs and outputs and the potential environmental impacts of a product system throughout its life cycle."

LCA is an objective process used to quantitatively evaluate the environmental burdens associated with a product, process, or activity throughout its entire life cycle from the cradle, through all the processes (called gates), to the grave. It utilizes various mass and energy balance protocols as well as environmental impact evaluation techniques to model the associated impacts across every stage of the life of a product. This may include the impacts associated with processing materials as well as the impacts associated with subsidiary actions such as waste disposal or beneficial reuse of processing materials. For example, in the extraction of natural resources, LCA includes the impacts associated with the extracted materials as well as the impacts associated with the extraction process.

Consider the cutting of trees to make wood products. The impacts associated with this activity include the environmental impacts from the loss of the trees as well as from the logging technology itself. Depending on the objective for which an LCA is performed, the range of the scope is as follows:

1. Raw materials extraction to the disposal of finished goods (i.e., cradle to grave)
2. Raw materials extraction to finished goods (i.e., cradle to gate)
3. One processing stage to another (i.e., gate to gate)

An LCA contains three phases: goal and scope definition, inventory analysis, and impact analysis, and each phase is followed by interpretation. The *goal* specifies the reasons for carrying out the LCA. The goal is important because the parameters to be used in the assessment are usually dependent on what the intended objectives are. The goal also specifies the intended application of the LCA as well as the intended audience. The *scope* helps to establish the system boundaries and the limits of the LCA. For example, if an LCA is to be performed to compare products, it is usually not the products themselves that are the basis of the comparison but rather the functions they provide. As such, a useful term, called the *functional unit*, is defined in the scope.

The functional unit is used to establish a basis for comparison of two products by identifying a common function and how each product achieves that function over its life. The functional unit is quantitative and measurable, and a careful and proper identification of the functional unit is crucial, particularly if the LCA is used for comparisons. For example, if an LCA is to be performed comparing reusable plastic cups and disposable paper cups, it would be incorrect to compare the impacts associated with one paper cup and those associated with one plastic cup since the plastic cup will continue to perform its function after the paper cup is discarded. Rather, it is more appropriate to determine how many times one plastic cup will be used before disposal and to calculate the equivalent number of paper cups needed to fulfill the same function; the comparison will be of the impacts associated with one plastic cup and the functional equivalent number of paper cups.

Another example involves comparing the life cycle impact of incandescent bulbs and compact fluorescent lamps. One fluorescent lamp uses significantly less energy than an incandescent bulb to produce the same amount of visible light. Since the function of the incandescent bulb and the fluorescent lamp is to produce light, it would be incorrect to compare a 40-W incandescent bulb to a 40-W fluorescent lamp. A 9-W fluorescent lamp produces approximately the same amount of light as a 40-W bulb. The functional unit would therefore be the amount of visible light required, and the LCA would be performed comparing the number of incandescent lightbulbs and compact fluorescent lamps required to provide the required amount of light.

However, a fluorescent lamp also lasts significantly longer than an incandescent bulb, so one must use multiple incandescent bulbs over the lifetime of one fluorescent lamp. Incandescent lightbulbs are typically rated to last an average of 1,000 hours, while compact fluorescent lamps are rated to last up to 8,000 hours. This means that, in addition to comparing the two choices based on the amount of light that is needed, the functional unit will also include the length of time that the light will be provided.

Inventory analysis involves determining the quantitative values of the materials and the energy inputs and outputs of all process stages within the life cycle. This includes the following:

- Raw materials/energy needs
- Manufacturing processes
- Transportation, storage, and distribution requirements
- Use and reuse
- Recycle and end-of-life scenarios such as incineration and landfilling

Inventory analysis is usually initiated with a flowchart or process tree identifying the relevant stages and their interrelations. Relevant materials and energy data are then collected for each process stage using materials and energy balance protocols to account for unknown values. Standardized units are used to make calculations and comparisons easy. It is also in this phase that the system boundaries necessary to meet the predetermined goal and scope of the LCA are determined. The *system boundaries* are the limits placed on data collection. For example, if the goal of the LCA is a comparative study of two products for which some life cycle stages are the same with the same materials and energy inputs and outputs, then the system boundary may be drawn to exclude the data related to those stages. Including these would only burden the LCA without providing additional information.

Impact assessment entails determining the environmental relevance of all the inputs and outputs of each stage in the life cycle. This includes the environmental impacts associated with the production, use, and disposal of the products. Relevant impact categories are selected, such as degradation of ecological systems, depletion of natural resources, and impacts on human health and welfare. For example, if we conducted the LCA of paper-based textbooks, examples of ecological systems degradation can include the impact of cutting trees as well as the impact of chemicals discharged during pulping on water quality. Trees are natural resources and renewable but not replaceable rapidly enough to immediately offset the impact of logging. The impact of the chemicals used on the health of the populations directly connected to the paper processing industries will also be assessed.

LCA results present the opportunities to reduce or mitigate identified environmental impacts arising from the manufacture, use, and disposal of products. The product design may be changed, the materials used may be replaced with less impactful materials, and the entire industrial process may even be changed based on the results of an LCA. The LCA represents the most objective tool currently available to inform decisions on the environmental sustainability of products and processes.

Emissions to the environment and the extraction and use of natural resources are called environmental stresses. What are the consequences of the emissions and the use of resources that are quantified in the inventory analysis? The ***impact assessment*** phase translates these into the relevant impact categories. We do this because the actual impacts categories are easier to relate to and understand. Impact assessment also makes results more comparable to guide decision making. For example, several gases contribute to the greenhouse effect. Simply knowing the quantities of these gases that may be emitted from a system is not sufficient, but evaluating how these emissions contribute to global temperature increase, climate change, sea-level rise, biodiversity changes, and food availability makes it easier to address specific concerns.

LCA is a useful tool for evaluating the comprehensive impacts from the manufacture, use, and disposal of products. LCA techniques rely on compiling databases of relevant energy, material inputs, and environmental impacts; evaluating potential impacts associated with process waste; and providing a simplified scale for interpreting modeling results. The results of an LCA can illustrate comparative differences in environmental and health impacts, such as whether product A or product B has a greater likely impact on climate change.

There are, however, significant ***limitations*** to LCA results that the user must recognize and interpret. Comprehensive LCAs require large amounts of resources and time. Accurate data collection is central to a reliable assessment, and the value of an LCA is only as good as the data used. LCAs are usually performed with truncated boundaries to limit the amount of extraneous data, implying a compromise for practicality. While LCAs offer insights into the environmental performance of products, they do not provide information on cost effectiveness, product efficiency, or any social considerations. LCAs produce good global-scale impact results, though assessments of local and contextual impacts are still challenging. In general, LCAs are decision support tools that are used alongside other points of consideration. The simplified LCA modeling results are suitable for relative comparisons of environmental and health impacts, but they do not indicate actual measurable responses, or the risk levels associated with the environmental and health risks.

The U.S. EPA and ISO standard 14043 recommend three key steps in **interpreting** the results of an LCA:

1. Identification of the significant issues based on the LCA
2. Evaluations that consider completeness, sensitivity, and consistency checks
3. Conclusions, recommendations, and reporting

Due to the large quantities of data that have to be collected and processed for any LCA, computer software is usually used to perform most LCAs. Many institutions and companies have developed their own software using data specific to their needs and concerns. There are, however, commercially available software that utilize comprehensive databases of materials, energy, and emissions inventories as well as multiple impact assessment methods. Some software are used to simply model the *Inventory Analysis* phase of the LCA, while others are useful for more complete LCAs that include *Impact Assessment* and *Interpretation* of results.

Part 5: Design for Sustainability

8.21 Determine if there are any existing synergies in the production process that could utilize the waste from one process as a resource in another process and show these on your diagram.

8.22 Brainstorm, identify, and list one or more industries that could use waste from a fly rod manufacturing process as a raw material or resource in producing a different process.

8.23 There are many options and many marketable designs for a fly rod—thus the reason there are so many different manufacturing companies and rod types within a company's line of fly rods. This is your time to ask questions and push boundaries. It is okay to realize you made a mistake later after the analysis. Can you innovate and conceptualize a design no one else has considered? Think of an alternative material for the components in your fly rod design that may be "greener." It *may* be bio-based, locally sourced, and so on. Find one such alternative for each of the following component options:

a) Rod blank
b) Handle
c) Reel seat insert and hardware
d) Fly line guides
e) Guide wrapping thread
f) Protective coating to cover the guide wraps

8.24 Update your previous design with any design changes you may have made in your response to Problem 8.20.

a) Briefly describe the materials (a)–(f) for each of the components listed above.
b) Cite a possible source for the new materials chosen.
c) Describe the cost per rod for each choice of materials (a)–(f).
d) Update the sketch of the material flow process from cradle to grave (or cradle to cradle) for each material choice.

Part 6: Simple Carbon Footprint LCA

8.25 The global fishing rod market was valued at $1.03 billion in 2021! It is estimated that about 500,000 fly-fishing rods are built each year. You will create a company to produce environmentally friendly fly rods. You will base your marketing materials on an assessment that includes embodied energy analysis and an analysis of greenhouse gas emissions. Write a short paragraph describing the goal and scope of an LCA to estimate greenhouse gases associated with your manufacturing process. This assessment will be useful to venture capitalists to determine if they should invest in your company.

8.26 Create an inventory analysis of the materials required for one full year of operation for your company.

a) Estimate how many rods your team will produce in a year and offer for sale as part of a proposal for a business plan (this may range from tens of rods to thousands of rods, depending on your customers).

b) Determine and report the total mass of raw materials needed for each component used to build your rod. (Don't forget to include estimated waste produced in each step.)

c) Find a source for the materials used in your rod. (This may be the raw material extracted or where a finished component is produced before it goes to your assembly plant.) Look online for a source material or supplier for each *component*.

d) Determine the distance from the source of each material through your process from raw materials to manufacturing and then on to a supplier near you. Make any appropriate assumptions that are necessary.

8.27 Show the boundaries of your *material and energy flow analysis* from Problem 8.24.

8.28 Note any important assumptions and considerations you will make to begin the greenhouse gas analysis. Note: This response may evolve over time and change as you perform the analysis.

8.29 Describe the expected limitations required to begin the greenhouse gas analysis.

8.30 Determine the impact your company would have on greenhouse gas emissions. Use Table 8.2 to estimate CO_2 emissions associated with the transportation for a year's worth of fly-fishing rods. At a minimum, include an analysis for the following:

a) Transportation by truck
b) Transportation by boat

Table 8.2 Carbon footprint of common materials

Material	Embodied Carbon (kg-CO_2/kg)	Material	Embodied Carbon (kg-CO_2/kg)
Aluminum—general[1]	8.24	Plastic—polycarbonate[1]	6.03
Aluminum—recycled[1]	1.69	Plastic—polypropylene[1]	4.98
Aluminum—virgin[1]	11.46	Plastic—polystyrene[1]	2.71
Brass[1]	2.64	Plastic—polyurethane[1]	3.76
Carbon (graphite) fiber composite[3]	10.1	Polyester[4]	6.4
Ceramics[1]	0.70	PVC[1]	2.61
Copper—tube and sheet[1]	2.71	Resin—epoxy[1]	5.70
Cork insulation[1]	0.19	Resin—melamine[1]	4.19
Cotton[4]	8.3	Rubber[1]	2.66
Freight[4]—air	4.42 per 1,000 km	Silk[4]	7.63
Freight[4]—boat/ship	0.03 per 1,000 km	Steel—general[1]	1.90
Freight[4]—rail	0.05 per 1,000 km	Steel-galvanized sheet—recycled[1]	0.68
Freight[4]—truck	0.21 per 1,000 km	Steel-galvanized sheet—virgin[1]	2.84
Glass[1]	0.91	Steel—recycled wire[2]	2.83
Glass-reinforced plastic (GRP) fiberglass[1]	8.10	Steel—stainless[1]	6.15
Nylon[4]	7.31		
Paint—general[1]	2.91	Water[1]	0.001
		Wax—paraffin[4]	0.609
Paint—solventborne[1]	3.76		
Paint—waterborne[1]	2.54	Wood—glue-laminated timber[1]	0.84
Plastic—general[1]	2.73	Wood—kiln-dried hardwood[1]	0.30
Plastic—ABS[1]	3.05	Wood—laminated veneer lumber[1]	0.63
Plastic—HDPE[1]	1.57	Wood—particleboard[1]	0.84
Plastic—LDPE[1]	1.69		

Sources: Hammond and Jones (2011)[1]; Lawson (1996)[2]; Kara and Manmek (2009)[3]; CO_2 Everything[4].

8.31 Create a visual aid to illustrate the impacts of the transportation of materials in your rod-production process. Use the output to show your results to create a visual illustration of the emissions of CO_2 and the embodied energy of the fishing rod.

8.32 The most important part of this assignment is to accurately portray the results of your analysis. This is what your grade will be based on (not whether your rod is more "environmentally friendly" than another team's design). Create a short video presentation. Each member of your group should contribute to the presentation graphics and narrative (voice). Also, feel free to use and build on previous submissions and data collected earlier to tell the story of sustainable design.

a) Describe the **goal and scope** of your analysis. What are the start and end points of your analysis? What are you hoping your analysis will illustrate?
b) Provide a **visual overview** of your design.
c) Illustrate the pathways of **energy flow and material flow** for your design.
d) Report the major **embodied energy** components associated with your design.
e) Illustrate the **steps modeled** in your assessment.
f) Provide an **interpretation** of the impacts associated with your design.
g) Describe the **limitations** and important assumptions in your analysis.
h) Describe the **benefits (economic, environmental, and social)** of your design and describe how these have been addressed using evidence summarized above.
i) List any appropriate **references**.

References

CO_2 Everything. *The Carbon Footprint of Everyday Products and Activities.* www.co2everything.com, accessed September 5, 2022.

Hammond, G., and Jones, C. (2011). *Embodied Carbon: Inventory of Carbon and Energy (ICE).* Prepared by the University of Bath, Bath, UK.

Kara, S., and Manmek, S. (2009). *Composites: Calculating Their Embodied Energy.* Sydney: University of New South Wales.

Lawson, B. (1996). *Building Materials, Energy, and the Environment: Towards Ecological Sustainable Development.* Canberra: RAIA.

Onkar, K. (2019). "The Environmental Impacts of Surf Wax." Master's Thesis, University of Queensland.

Index